GENETIC
ENGINEERING
OPPOSING VIEWPOINTS ®

Other Books of Related Interest

Opposing Viewpoints Series

Abortion
Animal Rights
Biomedical Ethics
Criminal Justice
The Environmental Crisis
Euthanasia
Global Resources
Health Care in America
Science & Religion

Current Controversies Series

Ethics
Hunger
Reproductive Technologies

At Issue Series

Business Ethics

GENETIC ENGINEERING

OPPOSING VIEWPOINTS ®

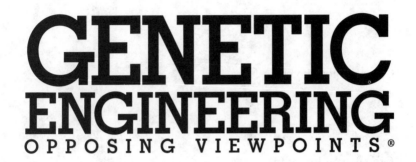

David Bender & Bruno Leone, *Series Editors*

Carol Wekesser, *Book Editor*

OPPOSING
VIEWPOINTS ®
SERIES

Greenhaven Press, Inc., San Diego, CA

Cover photo: Photodisc

Greenhaven Press, Inc.
PO Box 289009
San Diego, CA 92198-9009

Library of Congress Cataloging-in-Publication Data

Genetic engineering / Carol Wekesser, book editor.
  p.   cm. — (Opposing viewpoints series)
  Includes bibliographical references and index.
  ISBN 1-56510-359-9 (lib. ed. : alk. paper). — ISBN 1-56510-358-0 (pbk. : alk. paper)
  1. Genetic engineering. 2. Genetic engineering—Social aspects. 3. Genetic engineering—Moral and ethical aspects. I. Wekesser, Carol, 1963–   . II. Series.
QH442.G444   1996
174'.9574—dc20   95-241
                                                             CIP

"Congress shall make no law . . . abridging the freedom of speech, or of the press."

First Amendment to the U.S. Constitution

The basic foundation of our democracy is the First Amendment guarantee of freedom of expression. The Opposing Viewpoints Series is dedicated to the concept of this basic freedom and the idea that it is more important to practice it than to enshrine it.

Contents

Why Consider Opposing Viewpoints?

"*The only way in which a human being can make some approach to knowing the whole of a subject is by hearing what can be said about it by persons of every variety of opinion and studying all modes in which it can be looked at by every character of mind. No wise man ever acquired his wisdom in any mode but this.*"

John Stuart Mill

In our media-intensive culture it is not difficult to find differing opinions. Thousands of newspapers and magazines and dozens of radio and television talk shows resound with differing points of view. The difficulty lies in deciding which opinion to agree with and which "experts" seem the most credible. The more inundated we become with differing opinions and claims, the more essential it is to hone critical reading and thinking skills to evaluate these ideas. Opposing Viewpoints books address this problem directly by presenting stimulating debates that can be used to enhance and teach these skills. The varied opinions contained in each book examine many different aspects of a single issue. While examining these conveniently edited opposing views, readers can develop critical thinking skills such as the ability to compare and contrast authors' credibility, facts, argumentation styles, use of persuasive techniques, and other stylistic tools. In short, the Opposing Viewpoints Series is an ideal way to attain the higher-level thinking and reading skills so essential in a culture of diverse and contradictory opinions.

In addition to providing a tool for critical thinking, Opposing Viewpoints books challenge readers to question their own strongly held opinions and assumptions. Most people form their opinions on the basis of upbringing, peer pressure, and personal, cultural, or professional bias. By reading carefully balanced opposing views, readers must directly confront new ideas as well as the opinions of those with whom they disagree. This is not to simplistically argue that everyone who reads opposing views will—or should—change his or her opinion. Instead, the series enhances readers' depth of understanding of their own views by encouraging confrontation with opposing ideas. Careful examination of others' views can lead to the readers' understanding of the logical inconsistencies in their own opinions, perspective on why they hold an opinion, and the consideration of the possibility that their opinion requires further evaluation.

Evaluating Other Opinions

To ensure that this type of examination occurs, Opposing Viewpoints books present all types of opinions. Prominent spokespeople on different sides of each issue as well as well-known professionals from many disciplines challenge the reader. An additional goal of the series is to provide a forum for other, less known, or even unpopular viewpoints. The opinion of an ordinary person who has had to make the decision to cut off life support from a terminally ill relative, for example, may be just as valuable and provide just as much insight as a medical ethicist's professional opinion. The editors have two additional purposes in including these less known views. One, the editors encourage readers to respect others' opinions—even when not enhanced by professional credibility. It is only by reading or listening to and objectively evaluating others' ideas that one can determine whether they are worthy of consideration. Two, the inclusion of such viewpoints encourages the important critical thinking skill of objectively evaluating an author's credentials and bias. This evaluation will illuminate an author's reasons for taking a particular stance on an issue and will aid in readers' evaluation of the author's ideas.

As series editors of the Opposing Viewpoints Series, it is our hope that these books will give readers a deeper understanding of the issues debated and an appreciation of the complexity of even seemingly simple issues when good and honest people disagree. This awareness is particularly important in a democratic society such as ours in which people enter into public debate to determine the common good. Those with whom one disagrees should not be regarded as enemies but rather as people whose views deserve careful examination and may shed light on one's own.

Thomas Jefferson once said that "difference of opinion leads to inquiry, and inquiry to truth." Jefferson, a broadly educated man, argued that "if a nation expects to be ignorant and free . . . it expects what never was and never will be." As individuals and as a nation, it is imperative that we consider the opinions of others and examine them with skill and discernment. The Opposing Viewpoints Series is intended to help readers achieve this goal.

David L. Bender & Bruno Leone,
Series Editors

Introduction

"Genetic engineering poses moral and social dilemmas every bit as daunting as the rewards are enticing."

Stephen P. Stich

Genetic engineering, also called biotechnology, is a science in which genetic material is altered and manipulated in various ways in order to modify the characteristics of an organism or of a population of organisms. This new technology, which emerged in the 1970s as an offshoot of genetics, the study of heredity, has been applied to plants, animals, and humans.

In agriculture, genetic engineering is often equated with crossbreeding. For thousands of years humans have crossbred plants and animals to develop improved varieties and breeds. Crossbreeding allows farmers to weed out a plant or animal species' unwanted characteristics while retaining the desired ones. For example, by crossbreeding a hearty but bland-tasting variety of corn with a flavorful but less hearty variety, scientists might develop corn that is both hearty and tasty.

But genetic engineering is more sophisticated than crossbreeding. Combining two breeds is a slow, unpredictable process; and only plants in the same or a closely related species can be crossbred. Genetic engineering gives researchers more control, allowing them to insert and remove only the genes that regulate particular traits. In addition, scientists can now transfer a characteristic from one species to another. For example, the Arctic flounder possesses a gene that keeps it from freezing in icy water. Scientists are attempting to insert this "antifreeze" gene into a strawberry plant in order to create frost-resistant strawberries.

Along with agricultural applications, knowledge of genetics has also brought advances in medicine. Medical researchers are attempting to cure numerous health conditions by means of "gene therapy"—treatment aimed at correcting genetic defects. The first successful use of genetic engineering on humans occurred in 1990, when doctors used gene therapy to treat two girls suffering from an immunodeficiency disease. Nine-year-old Cynthia

Cutshall and four-year-old Ashanthi DeSilva both had a defective gene that did not produce a specific enzyme necessary for the correct functioning of their immune systems. Doctors inserted healthy copies of the gene into the girls' bloodstreams, thereby stimulating the production of the vital enzyme. Both girls soon became healthy, with fully functioning immune systems.

Dr. W. French Anderson, one of the three physicians involved in the procedure, called the girls' treatment "a social and cultural victory. It launched the field of human gene therapy." Many scientists echoed Anderson's enthusiasm and embraced genetic research as a revolutionary development: "The promise of diagnosing, preventing, and treating disease that will emerge from genetic research is considerable," maintains Arno G. Motulsky, professor of medicine and genetics at the University of Washington in Seattle.

But while the Cutshall and DeSilva cases may have "launched the field of human gene therapy," they also launched increased debate about ethical, medical, and social consequences of genetic engineering. While some contend that technology holds enormous promise for humankind, others believe it holds peril.

For example, enthusiasm about the potential of this technology to improve agriculture is tempered by fears among many who believe that altering the genetic structure of organisms could prove harmful to humans and the environment. For example, biotech companies are attempting to use genetic engineering to create herbicide-tolerant crops, according to writer Joel Keehn, in the hopes that they can "concoct a plant that can be doused with herbicide powerful enough to kill all the weeds growing nearby but won't bother the crop itself." According to Keehn, not only will the resulting use of large doses of herbicides threaten human health, but the herbicide-tolerant crops "could cross-pollinate with their wild, weedy relatives and give rise to herbicide-tolerant weeds—adding a new and particularly troublesome nightmare to farmers."

Others worry about the misuse of genetic engineering on humans. They fear that governments or others in power might force people to be genetically altered or might discriminate against people believed to be genetically "inferior." As writer Erik Lindala states, "The dark spectre of eugenics is inevitably raised when science anticipates the ability to change everything from physical traits to personality and behavior." Arthur L. Caplan, director of the University of Pennsylvania's Center for Bioethics, argues that these concerns should encourage the thoughtful pursuance of genetic engineering technology: "I would rather have genetic knowledge in my bag to fight disease than not to have it. But we need to be wary about the desire to enhance ourselves, to improve ourselves."

Genetic engineering has the power to alter life itself. The potential of this power to benefit and harm humankind is a theme that runs through many of the debates in *Genetic Engineering: Opposing Viewpoints*, which contains the following chapters: Does Genetic Engineering Benefit or Harm Society? How Does Genetic Engineering Affect Agriculture? Is DNA Fingerprinting Accurate? How Will Genetic Engineering Affect Health Care? How Should Genetic Engineering Be Regulated? The contributors to this anthology present a wide variety of opinions on the promise and peril of genetic engineering.

Does Genetic Engineering Benefit or Harm Society?

**GENETIC
ENGINEERING**

Chapter Preface

Genetic engineering offers humankind the potential of eradicating many diseases and of dramatically improving agriculture. But many people fear that this new technology may also threaten society in real and terrifying ways.

Some envision horror movie–like scenarios involving mutant creatures and environmental chaos. Others, recalling the Holocaust, fear the emergence of a eugenics movement bent on altering the human race to conform with its own idea of perfection.

While many experts reject these fears as exaggerated, others take them seriously but offer reassurances that genetic engineering, if responsibly conducted, will bring more benefits than threats to human progress. The contributors to the following chapter debate the effect genetic engineering might have on society.

"Genetic engineering is to traditional crossbreeding what the nuclear bomb was to the sword."

Genetic Engineering Is Dangerous

B. Julie Johnson

Genetic engineering poses many risks to humans and animals, B. Julie Johnson argues in the following viewpoint. Johnson describes genetic engineering experiments that produce monstrous, deformed, suffering animals. She believes such experiments are inhumane and dangerous. Johnson fears that genetic engineering will result in the devaluing of life because genes, embryos, and ultimately people will all be thought of as commodities to be bought and sold. Johnson is an ecofeminist writer and activist.

As you read, consider the following questions:

1. What is transgenic research, according to the author?
2. According to Johnson, what double standard was revealed by the government's decision on animal patents?
3. Why will genetic engineering especially harm women, according to the author?

B. Julie Johnson, "Patenting Life," *Ms.*, November/December 1992. Reprinted by permission of *Ms.* magazine, ©1992.

History repeats itself. In 1962 in *Silent Spring*, Rachel Carson issued a warning about the agrichemical industry's widespread use of synthetic pesticides. The industry marketed these chemicals with a pretense of altruism, as the end of crop damage, the control of nature's "enemies," and the panacea for world hunger. Today, however, agricultural pests have only increased; health officials have recognized hunger as a political, not a nutritional problem; and our soil, groundwater, and rivers are contaminated with poisons. All of this because a handful of technocrats saw an opportunity for profit. These men openly ridiculed Carson as a spinster, an animal lover, and a nature fanatic.

Today, a new generation of technocrats threatens to endanger the health of our planet once again. This time they want not only to control nature but to "improve" it. In genetic engineering, scientists manipulate deoxyribonucleic acid (DNA), the substance that encodes genetic information passed from one generation to the next. Experimenters can now cross species boundaries in ways previously unimagined, inserting, for example, human DNA into pigs, and cattle DNA into fish. In fact, they have even crossed kingdom boundaries: when the light-emitting gene from a firefly was inserted into a tobacco plant, the leaves and stem expressed light. Recent government decisions have actually awarded scientists patents on genetically engineered life forms as if they were "inventions."

Defective Animals

In 1987, Dr. John Hasler, cofounder of an animal biotechnology company in Pennsylvania, predicted, "We're going to make animals that nature never made." In fact, transgenic research (transferring genes from one species to another) is based on the theoretical potential to create "custom-designed" animals for food production, drug production, environmental cleanup, and disease "models." For example, experimenters at the U.S. Department of Agriculture inserted the gene that produces human growth hormone into pigs in order to create animals that would grow larger and faster than their normal siblings, on less food. Instead, the pigs were crippled with arthritis and had gastric ulcers, enlarged hearts, dermatitis, and kidney problems. Unfortunately, these transgenic beings were also somewhat leaner than pigs without the human gene. The experiments, therefore, continue despite the animal suffering and the unknown health effects of eating meat with a genetically altered fat content. This experiment demonstrates the unpredictability of genetic engineering, in spite of the biotechnocrats' myth that they are fully in control of DNA manipulation.

The concept of genetically "improving" nature is not limited to nonhuman species. Reproductive biotechnologists whose work fo-

18

cuses on women and embryos have already begun to envision the enhancement of humans via the elimination of genetic "defects." Physicist Richard Seed first practiced reproductive technologies on female cattle and then cofounded a research enterprise for embryo transfer experiments on female humans. According to feminist writer Gena Corea, Seed stated that genetic manipulation of embryos will "start therapeutically" and later be used in controlling evolution. He observed, "Just generally trying to improve the human race is a good thing." His partner and brother, Randolph, a surgeon, has identified asthma, for example, as a genetic defect.

While critics allow that biotechnology has some positive uses, it now lies in the hands of decision-makers whose record and motivations are highly suspect. It is ominous that at this historically significant moment, when women face the continuing erosion of our reproductive rights, biotechnocrats—predominantly male—are maneuvering into a position of increased control over our reproductive functions. This control ranges from techniques like in vitro fertilization and embryo transfer experimentation to designing artificial wombs.

Nowhere is the profit motive underlying the biotech revolution more evident than in the patenting of human gene fragments and whole animals. Patents issued by the U.S. government grant inventors the right to exclude others from making, using, or selling their inventions within the United States and its territories for 17 years. By law, naturally occurring organisms are not patentable. However, in April 1988, the U.S. Patent and Trademark Office (USPTO) declared that a mouse in whom experimenters inserted a gene associated with cancer was a "manmade invention." The experimenters were responsible for 1 out of a possible 50,000 to 100,000 genes, but for the sake of commercial exploitation they called the "oncomouse" their "discovery." The DuPont Corporation, which supported the research, now has the license to market this animal—in fact any mammals (except humans) in whom the gene is inserted during the embryonic stage, as well as their offspring—just as any other commodities are marketed.

The Double Standard of Biotechnocrats

The landmark mouse patent—reflecting a decision made by a single administrative office, USPTO, behind closed doors— caused so much controversy that several bills for a moratorium on animal patents have since been unsuccessfully introduced. Meanwhile, in May 1992, Gen-Pharm International, a Silicon Valley biotech company, announced that USPTO would soon grant it the world's second and third patents on mammals. One of these patents is for a genetically modified mouse that lacks part of its immune system; the second for a mouse genetically

19

programmed to be more susceptible to cancer. More than 150 additional applications for animal patents are currently pending.

The patenting decision reveals a curious, though typical, double standard: when biotechnocrats wish to address public fears about transgenic research, they refer to genetic alterations as "minor" and compare them to traditional crossbreeding, "classical methods" practiced for centuries. However, when it comes to proprietary interests, the same genetic alterations become so major as to designate an entire animal—something no scientist has ever created—an invention.

Life as Property

Genetic technology is already shoring up the mega-multinational corporations and consolidating and centralizing agribusiness. Corporate giants like General Electric, Du Pont, Upjohn, Ciba-Geigy, Monsanto, and Dow Chemical have multi-billion dollar investments in genetic engineering technology. It is becoming increasingly clear that we are placing the well-being of the planet and all its inhabitants in the hands of a technological elite. Our scientists, corporations and military are playing with, and may eventually own, our genes. . . .

We must remember that the mind that views animals as pieces of coded genetic information to be manipulated and exploited at will is the mind that would view human beings in a similar way.

Carol Grunewald, *New Internationalist*, January 1991.

The U.S. government has defended its increasingly lax regulation of the biotech industry by using this comparison between genetic engineering and traditional crossbreeding. In response, Andrew Kimbrell, attorney for the Foundation on Economic Trends, states, "Genetic engineering is to traditional crossbreeding what the nuclear bomb was to the sword."

A wide array of groups has united to contest the patenting of life—groups that seldom unite on other issues. These include religious organizations concerned about the sanctity of life, environmentalists worried about the release of genetically altered organisms, farmers anxious about paying patent royalties for their animals, scientists troubled about the free flow of information, and animal protectionists concerned about the increased exploitation and suffering of animals.

It is no accident that the first animal patents have been granted under the aegis of medical research. Although the most lucrative market for animal patents will probably be that of animals used for food production, it is more publicly acceptable to begin patenting complex life forms via mice that offer "miracle cures" rather than pigs that produce leaner meat. The pattern is familiar. Recall physicist Richard Seed's statement: "It [gene manipulation of human embryos] will start therapeutically."

The Commercialization of Reproduction

The National Institutes of Health has taken the profit motive one step further and applied for patent rights on human DNA fragments. These fragments are not complete genes but gene tags, whose function is not yet even known. After these applications were filed, Senator Mark O. Hatfield (R.-Oreg.) called for increased congressional oversight of patenting, warning that "careful examination has not taken place in the case of the genetic alteration and patenting of human genes and body parts, or in the case of the creation and patenting of transgenic animals."

Since the U.S. policy allowing the patenting of genetically modified animals has set the course for possibly patenting human gene parts, this policy may also be used in defense of patenting genetically modified human embryos. Unless the decision to patent animals is rescinded, we can also expect increasingly blatant commercialization of women, our reproductive functions, and embryos.

Already women have been totally objectified in reproductive biotechnology. We have been referred to as "surrogate uteri" and "alternative reproduction vehicles." Most victimized by further commercialization of human reproductive procedures will be those most vulnerable, such as poor and Third World women, who have already been identified by at least one surrogate broker as a source of wombs.

The European Patent Office (EPO) has granted a patent to the Howard Florey Institute of Experimental Physiology and Medicine in Melbourne, Australia, for a gene found in the ovarian tissue of pregnant women. The gene codes for the human hormone relaxin, which facilitates childbirth by relaxing, or softening, the cervix and birth canal. The Greens in the European Parliament have filed a legal claim with EPO on the grounds that the gene is not an "invention," but exists naturally in humans. They have expressed concern about the medical commoditization of women's bodies.

As H. Patricia Hynes notes in *The Recurring Silent Spring*, environmentalists and animal rights activists who warn about biotechnology are targeted in the same way as was Rachel Carson: "eco-cranks," "anti-intellectuals," "the new Luddites," "animal lovers," and "nature fanatics." The major difference between the agrichemical industry of the 1950s and 1960s and the biotech revolution now is this: biotechnocrats possess the tools to change the course of evolution.

"Genetic engineering is not so dangerous after all."

Genetic Engineering Is Safe and Beneficial

Bernard D. Davis

The late Bernard D. Davis was Adele Lehman Professor of Bacterial Physiology at Harvard Medical School in Cambridge, Massachusetts. He edited the book *The Genetic Revolution: Scientific Prospects and Public Perceptions*. In the following viewpoint, Davis outlines many benefits of genetic engineering, from medical advances to improved crops. He examines the public's fears concerning genetic engineering, and responds to these fears by explaining his view that genetic engineering is safe and controllable.

As you read, consider the following questions:

1. How can science predict the effects of genetic engineering on society, according to Davis?
2. Why should the public not fear that genetic engineering will create epidemics, according to the author?
3. How have the courts and public agencies affected the public's perception of genetic engineering, according to the author?

Excerpted from Bernard D. Davis, "Genetic Engineering: The Making of Monsters?" Reprinted with permission of *The Public Interest*, no. 110, Winter 1993, pp. 63-76, ©1993, National Affairs, Inc.

In 1973 scientists integrated a number of esoteric techniques in microbial and molecular biology, making possible the directed molecular recombination of DNA. By this method, fragments of DNA from any source could be spliced in the test tube and cloned in host organisms. Scientists soon devised other ingenious techniques for manipulating DNA, including improved methods for isolating genes and determining their sequences. These developments have had a major impact on research in virtually every branch of the biomedical sciences. They have also created a burgeoning biotechnology industry that encompasses medicine, agriculture, and pollution control.

But despite the outstanding achievements and promise of this genetic revolution, the public has been ambivalent. People are eager for the benefits but fear the possible dangers. By now, after twenty years of expanding experience with biotechnology with no detectable harm to humans or to the environment, the anxiety has abated a good deal. Nevertheless, the development of safety regulations for bioengineering is still plagued by confusion, controversy, and continuing public apprehension. Why has there been such concern over essentially hypothetical dangers? . . .

Reasons for the Public's Uneasiness

I would first note a unique feature of this controversy: its prolonged focus on hypothetical risks, rather than on the more usual exaggeration of demonstrated ones. With newly recognized bona fide sources of harm, such as asbestos or radon, we ordinarily react slowly and then overreact after a lag. But with recombinant DNA we reacted explosively and we continue to debate the issues vigorously—even though the basis for predicting future harm from recombinants has become exceedingly tenuous (with the exception of products of organisms that are themselves pathogens, that is, sources of disease).

Concern about genetic engineering has been further intensified by an underlying uneasiness over the future impact of gene therapy and genetic screening in human beings. Here, there are clearly serious concerns. These include the potential tendency of therapeutic purposes to blur into eugenic ones, the likelihood that knowledge of individual susceptibilities to future disease will often generate more anxiety than benefit, and the certainty that such knowledge will greatly increase problems of privacy. These concerns have been most influential in those countries where distortions of genetics contributed to the Holocaust: The Green political party has impeded progress in genetic engineering in Germany even more than anti-technology activist Jeremy Rifkin has been able to do in the United States. Concern for environmental deterioration has led to similar reactions in other countries, and in Switzerland it has driven out some biotechnology-based industries.

Still another, more general reason for uneasiness over genetic engineering has been an extrapolation from the model of the physical technologies. A few decades ago the advances in these technologies seemed to be providing us with a virtually free lunch; but disillusion set in as we encountered unanticipated costs to our environment and our security. As a result, some fear that manipulating the cell nucleus might, like manipulating the atomic nucleus, have unforeseeable costs.

Natural Adaptation

The assumption behind these concerns is that we have no basis for estimating future dangers from biotechnology. In fact, we do have a basis for predicting the effects of biotechnology—our historical experience with domestication. Domestication began when our ancestors learned to tame certain animals, plants, and fermentation microbes to serve human needs, then discovered how to select empirically for varieties strengthened in certain valuable traits. The benefits have been strikingly free of social costs for thousands of years, in contrast to the more mixed bag yielded by the physical technologies.

Furthermore, the products of past domestication have not "taken over" via spontaneous spread, as is feared for the products of the new biotechnology. They have spread only to the extent that cultivation by humans has caused them to displace the earlier occupants of the same territory. Since the products of the new biotechnology are based on an extension and refinement of the same principles that govern domestication, they should be subject to the same limitations on their spread.

Better Crops Than Nature Can Produce

Genetic engineering makes it possible for scientists to add or eliminate specific traits from individual plants, animals, or microbes. . . . Genetically engineered, or *transgenic*, plants may have improved abilities to survive drought, frost, salinity, or certain pests. . . . Advances in the field of genetic engineering could mean progress on an unprecedented scale for all of civilization.

Gail Dutton, *The World & I*, August 1991.

These limitations arise from the nature and scale of the evolutionary process. Evolution has been continuously experimenting with genetic novelties for three billion years. It has been extraordinarily effective in filling each ecological niche with organisms exquisitely adapted to that environment, from the

Alaskan tundra to hot vents in the depths of the ocean. More-over, in the microbial world the scale of natural adaptation is enormous: The average shovelful of soil contains as many indi-vidual creatures as the total human population. By comparison, our genetic experiments in the laboratory are puny.

Accordingly, the likelihood that we can further improve on the adaptation of an organism to its natural environment is virtually nil. (We may be able to improve the adaptation of an organism to an artificial environment, such as a farm. But this advantage is limited to the boundaries of that environment.) When we breed for "improvement" in an organism—an increase in a prop-erty that serves us—both theory and empirical evidence point to a *decrease* in its adaptation to the environment from which the parental strain was taken. . . .

Limits to Genetic Novelty and Spread

But if we can now remake organisms at will, is there not a qualitative difference between modern genetic engineering and classical domestication? With an unlimited range of products, might not some inadvertently spread beyond our control?

Several arguments should allay this concern. First, even though we can indeed manipulate DNA in the test tube at will, it does not follow that we can modify organisms at will. In or-der for an organism to develop and function effectively its parts must interact in a coordinated manner, fitting each other like the parts of a smoothly functioning machine. Hence, only those new variants that have a sufficiently coherent, balanced set of genes can survive.

Furthermore, even if a radically altered organism *is* nominally viable, it suffers significant disadvantages in evolutionary compe-tition. Recombination of ill-matched genes from distant sources will yield poorly adapted organisms, not the vaguely conceived, dangerous monsters of current science fiction. . . .

I suspect that these rather theoretical arguments did not have much to do with the gradual realization that genetic engineering is not so dangerous after all. Most important was the simple ex-perience of expanding the work into thousands of laboratories without harm. Another factor was the eventual recognition that organisms containing foreign DNA arise in nature and hence are not as utterly novel as was initially assumed. The initial assump-tion of great novelty was understandable, not only because the technical advance was an enormous one, but because molecular biologists must have felt a Promethean pride in having appar-ently created combinations of genes that could never before have existed on earth. But scientists did not create *de novo* [anew] the several steps that they used in developing DNA splicing in the test tube. These steps all occur in nature, and the key discoveries simply improved their efficiency. So while these discoveries were essential for producing recombinant bacteria in the laboratory in great variety and in usable quantities, it is difficult to avoid the conclusion that in nature bacteria must also take up DNA from foreign sources and produce recombinants, though at a very low rate and with a very low survival value. . . .

One other feature of the microbial world provides further reassurance: the stringent requirements for pathogenicity. The vast majority of bacteria are not pathogens—they do not cause disease. They are found in soil and bodies of water, where they convert organic matter to simple degradation products (carbon dioxide, ammonia, etc.), which recycle into other microbes or into plants. One does not easily make an effective pathogen out of such harmless bacteria. For with pathogens, as with benign microorganisms, evolutionary success is not ensured by any single, powerful gene; it depends on an effectively interacting ensemble of genes.

Consider the example of diphtheria toxin. The gene that codes for this potent toxin is found in nature only in the diphtheria bacterium, where it is accompanied by other genes that help make the organism an effective pathogen. Diphtheria toxin is not found in any other bacterium. Hybrids that formed the toxin must surely have arisen in nature from time to time but did not survive.

Man-Made Epidemics?

With increasing recognition of these arguments against a special danger in recombinant bacteria, the guidelines for the use of such organisms, in research and in industry, were progressively relaxed. The issue seemed to be pretty well settled. But in 1984 a new wave of concern arose, as scientists began developing potential applications of biotechnology that would involve the deliberate introduction of engineered organisms into the environment. Examples include the use of such organisms to replace nitrogen in fertilizer, to replace toxic chemical pesticides, to digest toxic organic pollutants (such as oil spills), or to prevent frost damage on crops.

This latest alarm over biotechnology has centered largely on two existing models. The first is of damage to the environment from toxic chemicals. The extent of this damage depends on the scale of the introduction. There is an important difference, however, between this model and that of bioengineered bacteria. With chemicals the harm is created directly by the introduced material, while with bacteria the harm would depend on the uncontrolled multiplication of the progeny. Such spread in turn would depend on the ability of the introduced organism to compete, in a Darwinian world, with those organisms that are already present. And the effect of scale on that competition is clear. If an introduced soil organism is not competitive, even huge numbers can have only a transient and local effect before dying out. Conversely, if it should be more competitive than the native organisms (though that would not be expected, for reasons presented above), even a small, accidental escape from the laboratory could start a spread, just as a single import of smallpox can start an epidemic in a susceptible population.

Thus, scale of introduction is not decisive for competing bacteria. Of course, one can imagine that on a huge scale the "transient and local" effect could be significant, even though it would not result in spread. But that problem would not come as a surprise, and we should be able to control it.

Preventing and Curing Illness

Imagine beating chronic, debilitating, even fatal diseases before they strike. Think of the lives, the medical dollars, that could be saved if doctors could identify individuals genetically predisposed to heart disease, cancer, and other killers, and, through modification of diet, lifestyle, or other risk factors, reduce or eliminate their susceptibility. The possibility seems within reach as an ever-expanding arsenal of gene-testing technologies is developed.

Andrea Kott, *American Medical News*, August 22-29, 1994.

The second major basis for concern over introduced organisms is the harmful spread of certain "exotic" organisms after their importation from distant regions. This analogy has had widespread appeal. The unexpected and costly spread of certain imports, such as starlings and kudzu vine in the United States, or rabbits in Australia, has legitimately caused great concern to ecologists. But the parallel to engineered organisms is weak, and perhaps even specious, because of a key difference: One process moves an unchanged organism to a new environment, while the other changes the organism and then returns it to the original environment.

This distinction has large consequences. Specifically, non-engineered exotic transplants have already been well adapted by evolution to their native environment, where their population density has been limited by various physical and biological factors. In a new environment that lacks these factors the organisms will proliferate excessively. Engineered organisms, in contrast, are ordinarily returned to the *original* environment—and as we have already noted, they are highly likely to be *less* well adapted than the parental strain to that environment. (Of course, the altered organisms may be transplanted to a new environment; but then we would be dealing not with a special problem of recombinants but with the old problem of exotic pests.) Recognizing these considerations, many ecologists have stopped stressing the analogy to transplanted species. Its public appeal, however, persists.

There is still another source of reassurance: the extensive experience already accumulated through the use of genetically modified microbes in agriculture. Such organisms, obtained by traditional genetic methods, were introduced long before recombinants became available. They include strains of *Bacillus thuringiensis* (used to kill insect larvae) and nitrogen-fixing bacteria (used to spare the need for nitrogen in fertilizer). Their regulation was straightforward, and no harm has been detected. Clearly, commercial use of recombinant bacteria, as of any other bacteria, will require similar measures. But the purpose will be primarily to avoid toxicity for humans and animals, rather than to anticipate uncontrollable, harmful spread. (A parallel experience in medicine has been regulation of the use of live, attenuated viruses as vaccines, including smallpox, polio virus, mumps, measles, and rubella.) . . .

Prospects for the Future

It is easy to arouse public suspicion of microbes, which are most familiar as "germs" that cause disease. Past costs and errors in the exploitation of new technologies have generated a cadre of political activists who have become skillful in using the courts and the media to promote their position. Moreover, the professional concerns of ecologists encourage conservatism about changes in the environment, while various lay environmentalist organizations are even more conservative—and emotional. The genetic revolution thus calls for a great deal of education, in our schools and to the adult public, on the beneficent, essential role of most of the world's microbes. . . .

Unfortunately, the courts and administrative agencies have often shared or responded to scientifically unsound, distorted public perceptions. It is to be hoped that, with experience and familiarity, they will learn to apply common sense to the increasingly important science of genetic engineering.

"All of us must commit ourselves to winning the biotech race."

The United States Should Continue Investing in Biotechnology

Richard J. Mahoney

Genetic engineering has produced a revolution in biotechnology by creating high-quality plants and animals that are more disease resistant and more nutritious, Richard J. Mahoney maintains in the following viewpoint. While the United States is the world's leader in biotechnology, according to Mahoney, it is facing strong competition from Japan and other nations. He concludes that Americans must continue to invest in biotechnology if the nation is to prosper. Mahoney is chairman and chief executive officer of Monsanto Company, a chemical company based in St. Louis.

As you read, consider the following questions:

1. Mahoney cites many benefits of genetic engineering. List five of these benefits.
2. What hurdles are preventing the United States from continuing its role as world leader in biotechnology, in the author's opinion?
3. What are some of the ways Americans can work to promote investment in biotechnology, according to Mahoney?

From Richard J. Mahoney's speech "Biotechnology and U.S. Policy," delivered February 12, 1993, to the Executive Club of Chicago, as reprinted in *Vital Speeches of the Day*, April 15, 1993. Reprinted here by permission of the author.

My company, Monsanto, has made a significant bet on this new science we call biotechnology.

But it is more than a science.

It is more than an investment by one company.

And it is more than a potential industry.

Indeed—it offers the tantalizing possibility of driving the next long-term economic thrust for the United States.

In time, it offers progress undreamed of in health care.

It offers the promise of stable food supplies for Third World nations.

Imagine Doing It Right

Imagine the major food crops—corn, wheat, rice, soybeans—which can resist diseases—and resist pests—and create their own fertilizers—and resist extremes of weather.

Imagine potatoes containing more protein, and other vegetables and fruits which contain more nutrients, taste better and resist rot. Can you imagine tomatoes that actually taste like tomatoes?

Imagine what such food crops could mean for a world population which will double in less than 40 years.

Imagine a fundamental revolution in health care—with treatments and perhaps even cures for heart disease, arthritis, Alzheimer's, cancer and AIDS.

And longer term, imagine manufacturing processes for the chemical industry which produce no waste to pollute the environment—and new methods to clean up the air, water and land.

And this time—for this new industry—America did it right.

Government money was plowed into biotechnology through the National Institutes of Health and our superb universities.

Start-up venture capital was made available to entrepreneurs.

Our regulatory system—not known for speedy approvals—even approved some new, biotech-based pharmaceuticals—while in Europe, these sciences were stifled, indeed driven to the United States. Hoescht and BASF moved their research programs to America—after an unfriendly reception in Germany.

No—this time—in biotechnology—we did it right—in agriculture—and in health care and pharmaceuticals. This focus and exercise of the collective national will in biotechnology has been an incredible achievement—rarely if ever duplicated in peace-time.

And with these new biological sciences are coming high tech jobs—high tech jobs paying high tech salaries. And miracle products are coming—products for a good life, products which can control and perhaps even eradicate major threats to health and well being.

And biotechnology is bringing a resurgence of our economy,

with these new sciences estimated to be $100 billion strong by the year 2000, and that's just the beginning. . . .

And our nation will benefit from our wonderfully inventive research, and the "R" of the "R&D" [research and development] will "D" in this country for a change—and we will lead the world. And we will train Third World scientists—like Dr. Florence Wambugu of Kenya, working right now in a Monsanto lab in St. Louis—to apply these sciences to the staple crops of Africa and Asia, so that they fight crop disease and feed their people.

So—for biotechnology—not to worry!

Worry!

Creating Hurdles

Precisely at the moment when we are on the threshold of capitalizing on everything we've done in these new sciences, right at the point where success seems assured, the American pistol seems once again aimed to shoot itself in the foot—or at least try to. We are forming a circular firing squad—taking dead aim on biotechnology right in the center.

Right when we have the future within our grasp, we are poised to throw much of it away.

America is creating some of its own "Made in the Good 'Ol U.S. of A." hurdles. To borrow from biological terms, the American immune system has begun working overtime to reject this new invader called biotechnology—we have started our own inoculation campaign to make sure this upstart industry is contained—restricted—and perhaps even stillborn.

I'm known to be fond of telling my people at Monsanto that policy is what you do—not what you say.

If you keep doing something long enough, it must be your real policy—no matter what you declare publicly your policy to be. Let me cite a few examples of implicit policy on biotechnology that becomes explicit reality.

Based on what America does, it must be our policy to make sure investment money is constrained at home so that other nations can capitalize on our lead. It must be our policy to make sure that Japanese companies make as many inroads as possible into these new biological sciences—because that is what our investment and capital formation disincentives encourage. If we keep doing it, it must be our policy.

Japanese Ambitions

For a number of years, the Japanese have been interested in biotechnology. So interested, in fact, that MITI—the Japanese ministry of industry and trade—declared biotechnology a national priority. It has been reported they proclaimed a national goal that one percent of Gross Domestic Product be invested in

31

these new sciences.

They became interested because they saw the promise and they saw the implications. They turned their eyes to America because the science was here, and they could catch up on the cheap. They didn't do this by stealth or by unfair trading practices. They paid for it. They openly announced their intentions, and they moved just as openly. And our capital gains tax and other risk capital taxes encourage it and penalize American entrepreneurs, often causing start-up companies after first-round financing to look elsewhere—anywhere—for needed capital if it's not available here. And the Japanese have willingly obliged— so it must be our policy.

Those of you involved with the electronics industry will find this an all-too-familiar story. Let me cite just a few Japanese forays from recent issues of biotech newsletters.

• Chugai Pharmaceuticals of Tokyo invested $30 million in a Boston firm to develop new products for autoimmune diseases like rheumatoid arthritis.

• Otsuka Pharmaceuticals of Tokyo announced a licensing agreement with Cetus of Emeryville, California, for the exclusive rights to use and sell human therapeutic materials here in the United States.

• Cytel of San Diego announced a five-year agreement with Sumitumo Pharmaceuticals of Japan to develop drugs for cancer and other diseases.

• And Sumitumo has also licensed drug screening technology from GES Pharmaceuticals of Texas, following its purchase in 1991 of an equity position in GES.

It's a small sampling.

In fact, more than 200 contracts for technology or product licensing arrangements were made between Japanese and U.S. biotechnology ventures in the early 1980s—and more since.

The Public's Lack of Knowledge

I have no problem with foreign investment in the United States; it's healthy—indeed, international transfer of technology is increasingly a fact of life. But this is at the core of our basic R&D and our tax laws encourage it. It's probably not the major factor, and maybe it wouldn't keep all of our science benefits at home, but why don't we encourage capital formation here? We don't favor risk capital. It must be our policy.

And it must be our policy to ensure that any new technology will automatically frighten large segments of the public.

Our system of public education seems designed to discourage the public from understanding even the rudiments of science and *starves* the real educational needs of American children, *especially* in the sciences. And it's no surprise that the American public doesn't understand science and is subjected to and falls for the science scare of the month headline and is unable to sort our real risk from just plain ignorance.

Biotechnology sometimes gets put in a special class of risk by the public—open to exploitation by the growth and influence of anti-science organizations who can scare the public in the process.

A Brave New Era

Scientists have already made remarkable discoveries concerning human genetics. They can link flawed genes to particular illnesses, including some forms of cancer. We are promised to be on the verge of a brave new era in medicine, where serious disease will be cured by patching the flaws in our intricate genetic engines.

Jeff Elliott, *The World & I*, March 1995.

The handful of groups behind these attacks wield an influence all out of proportion to their numbers because few seem able to differentiate between facts and nonsense. How easy it was to organize chefs against Calgene's tomato! [Calgene created the Flavr Savr tomato, which is genetically engineered to retain freshness.] Food processors will be the losers if they don't resist this regulation by activist groups—and not the government regulators. Food processors will have to become their own Food and Drug Administrations if they start responding to every pressure group.

It's no wonder that the public gets confused, and questions the products of biotechnology. The public can be led to perceive a risk where none exists, and it's on that basis that we must learn to communicate with the public, understanding the public's concerns—and responding to those concerns. We have not done an outstanding job of communicating with the public—but we are learning—and we're learning fast.

Enhancing Crops, Increasing Productivity

One area where we're learning is in *plant* biotechnology.

American companies are far in the lead developing the science which will eventually mean low-cost, high quality foods. Drought-resistant crops. Third World food staples which can resist fungal diseases. Even tomatoes that taste good because ripening is controlled. Our first potential product at Monsanto will be insect-resistant cotton, which could sharply reduce the amount of insecticides used on cotton in America—a significant reduction because 40 percent of all U.S. insecticide use is on cotton.

In the year 2000, it will be possible to enhance virtually every crop—not by black magic, but by doing what seed breeders have been doing for centuries: putting desirable qualities into crops—this time, with biotechnology doing it in a few years instead of the decades it takes by genetic breeding. America—and indeed the world—has a large stake in R&D in biotechnology-based plant genetics.

Here's the usual litany of those opposed to plant biotechnology—it's a small minority but it's vocal.

Only the developers of the plants will benefit.

God put the plants here and we shouldn't tamper with them—even though we've been, as I said, breeding crops for centuries.

No one knows what will happen when these plants are introduced into nature—they could become the soybean that ate Centralia. The standard line for delay is "we're not opposed, but more study is needed"—now that's after *every* government study group authorized to do so *has* to say it's *ok* when they examine and approve the individual uses.

And there *are* significant regulatory hurdles. Like a traditional pharmaceutical or food ingredient, each product of biotechnology requires an extended regulatory approval period. Regulation is very appropriate, in our view, and it's needed not only to make prudent regulatory judgments but at least as important to assure the public that the products of this new technology are indeed safe for man and the environment.

But the federal agencies—EPA [Environmental Protection Agency] and FDA [Food and Drug Administration]—are badly underfunded, and too often micromanaged by Congress. So for this and some process reasons, clearance is none too swift—at best.

The time required for regulation—7 or 8 years to get the data and get it reviewed—must be weighed against the time dimension of capital. When the regulatory review system for new products is hampered and extended by inadequate funds, or an inefficient review process, the product developer and the public both will suffer. And when the anti-science people put the heat on, the process slows up—it doesn't stop it, it just slows it up.

Isn't that interesting—unemployment payments *are* U.S. policy; acceleration of new products to relieve unemployment is *not* U.S. policy. It's the American way, I suppose—and policy is what you do.

Improving Health Care

Health care is another major area of biotechnology research, the one which has already gotten some products out into the market. The new therapies from biotechnology hold enormous promise in cost-effective health care.

Pharmaceuticals are a major commercial target and a *major* opportunity for biotechnology to help society. The underfunded regulatory process is further complicated by calls for fairness and equity.

In an attempt to combat health care costs, some have moved in on pharmaceuticals. The attempt to control health care costs is indeed a noble one—but pharmaceuticals are the wrong target. They represent only 6 percent of total health care costs and they're probably the most cost-effective part of the equation. Ask any ulcer victim who has traded a $15,000, not-very-effective surgical procedure for $1,000 a year of Tagamet or Zantac to relieve the ulcers.

In fact, if you took *all* the profits away from pharmaceutical companies—*all* of it—you'd cut U.S. health care costs by eight tenths of one percent—and eliminate any possibility of R&D breakthroughs on debilitating and expensive illnesses.

Strange Logic

It seems unconscionable to some that pharmaceuticals must provide a return on investment—a return to cover all the great but failed attempts in the laboratory. Frankly, I'd rather have a *successful* industry work on the next wave of U.S. growth than an unsuccessful one. Most major pharmaceutical companies spend about 15 percent of sales on R&D—now that's even more than corn flakes or booze [companies do], which make about the same financial return as pharmaceuticals.

Fortunately, pharmaceuticals still hold widespread support for their performance in health care cost containment and quality of life improvements.

But there's strange logic by some: if it works, fix it.

It's their version of the American way. It's their version of appropriate U.S. policy.

I haven't mentioned product liability—an institution unique to America. Tort reform hasn't come yet—and I expect to eventually see a plaintiff filing suit claiming "genes in my tomato." Wait until start-up companies see what larger companies have been facing for years in tort cases. Monsanto can handle it—we have the staying power to deal with these obstacles and we've had to face them before. But can the many superb science-based start-up companies do it with their funding limitations? Maybe—maybe not.

I've selected only a few examples of the series of roadblocks we're placing in front of technology advancement. There are many, many others. Some are common to all industries—but they all affect biotechnology.

The biotech race can be won in this country—and I'm confident it will be—because it must be. Because the benefits go well beyond commercial gain to important societal advances. And the U.S. has done a good job of getting it launched.

But *must* it be hobbled and hampered at every turn? *Must* it run with the wind in its face—rather than at its back?

The True Villains

This is what we are doing to ourselves in America. . . . The Japanese and the other usual suspects we round up are not the villains—Americans are the villains. Implicit U.S. policy is the villain. Policy is what you do—not what you say.

Ladies and gentlemen, I am not [speaking] to depress you but to rouse you to action. . . .

Never before have we so sorely needed people who will say "I will."

I urge you to write to [your congresspersons] and make them make sense at the state level and in Washington.

Urge improvement in capital gains tax policy to free investment in innovation.

Urge product liability reform—especially punitive damages at the state and federal levels.

Urge a national policy of scientific literacy for our young people.

Urge the media to reject fringe views of science scares masquerading as media balance.

Urge funding for regulatory agencies to reduce product approval times.

Urge a sense of national pride that we have a successful pharmaceutical industry willing to invest heavily in R&D—leaders in the world—instead of bashing it because it's profitable.

Urge Congress to follow the historic instruction given to new doctors: First do no harm—help if possible—but first do no harm. . . .

All of us must commit ourselves to winning the biotech race . . . and help create the next great revolution in agriculture—and in health care—and in industry.

> "*The manipulative power of the new biotechnologies poses a threat to public health, to the quality of the food we eat, and to the biological integrity of life itself.*"

U.S. Investment in Biotechnology Is Dangerous

Brian Tokar

Supporters of biotechnology suggest that it can help eradicate world hunger and cure many diseases that plague humanity. These visions are false, Brian Tokar contends in the following viewpoint. In reality, he asserts, biotechnology and genetic engineering threaten to irreparably harm the environment and expose people to new health hazards. Rather than supporting this new science, Tokar advises, Americans should oppose it and educate themselves about the dangers biotechnology presents. Tokar, an activist for environmental and social issues, is a frequent contributor to *Z Magazine*. (The original article contains additional examples with which the author illustrates his points.)

As you read, consider the following questions:

1. What social problems does Tokar foresee if genetic screening becomes more commonplace?
2. According to the author, why is the advance of biotechnology probably unstoppable?

Abridged from Brian Tokar, "The False Promise of Biotechnology," *Z Magazine*, February 1992. Reprinted by permission of the author.

Will biotechnology be the new American growth industry of the 1990s? Will developments in biotech restore U.S. competitiveness, get us out of the recession, and restore boom times to parts of the country that have suffered recent setbacks in the electronics and "defense" industries? This is the line that is being pitched on behalf of the drug and agro-chemical industries, and which city and state officials across the United States are lining up to buy.

In New York City, leading officials . . . have endorsed Columbia University's scheme to build a major biotechnology research park in Harlem, on the site of the famous Audubon Ballroom. The Ballroom, which played an important part in the life (and death) of Malcolm X and countless other African-American activists and artists, has become the center of a struggle pitting Black Consciousness activists, Left Greens, and others against much of the city's liberal establishment. In Philadelphia, activists opposed the demolition of an historic university building for a biotechnology facility funded largely by the Defense Department. Officials in Chicago, San Francisco, and other major cities are also maneuvering to recruit biotech companies into their communities, and new research facilities have aroused opposition in university communities like Burlington, Vermont, and Athens, Ohio. Three sessions of the New England Governors' Conference had a major focus on biotechnology, with governors from across the political spectrum boasting about the futuristic new industry that was going to pull *their* state out of the latest regionally devastating recession.

Perpetuating a Myth

In 1989, I offered an overview of biotechnology, which is actually an assemblage of new technologies dedicated to the industrial simulation and manipulation of fundamental life processes. I described how genetic engineering, the most far-reaching of these technologies, amounted to an attempt to redesign life at the cellular level to better serve the needs of mass production and commercial exploitation. Far from "curing" hunger and disease, biotechnology has served to close off more ecologically and ethically sound lines of research and scientific inquiry. Biotechnology, I argued, discourages attention to underlying problems by perpetuating the myth that the inherent ecological limitations of a nature-denying society can simply be engineered out of existence.

The biotechnology industry has become far more aggressive, as companies built on their scientific know-how and futuristic potential have pushed to get their products onto the market. All of the major international pharmaceutical companies have invested heavily in biotechnology, often buying up the smaller, specialized companies that helped establish the field. Wall Street has been extremely cooperative, with biotech stock offerings in 1991 far exceeding precedents.

Reprinted by permission of Mickey Siporin.

There has also been a large shift in marketing strategy. Far-fetched claims about the future benefits of biotechnology's products are now tempered with efforts to allay people's concerns about the process of genetic engineering, which, we are now told, is merely an incremental advance over traditional "biotechnologies" of plant and animal breeding, brewing alcoholic beverages and the like. Biotechnology will even promote "sustainable agriculture" around the world by making food crops and animals more efficient, they say. It should not matter to us that this new technology seeks to overturn the inherent biological divisions between living species, and even between plants and animals, reducing everything in nature to objects for commercial manipulation. Forget that it will ultimately make people far more dependent upon the products of multinational chemical companies. We are to marvel at its successes and buy its products, without batting an eye at what Indian feminist and physicist Vandana Shiva calls the commodification of life itself. . . .

"Barnyard Biotechnology"

In the coming years, chemical companies want to flood the market with new, genetically engineered food products. Nearly

300 varieties of genetically altered plants have been approved for field testing in the U.S., according to an ongoing survey by the National Wildlife Federation's National Biotechnology Policy Center. Many are designed to resist common plant diseases or produce toxins that kill pests; the latter are expected to seriously impair the effectiveness of naturally applied biological controls, such as the bacterial sprays commonly used by organic farmers. Others are engineered to resist specific herbicides, so higher doses of herbicides can be used by growers to kill weeds. Tomatoes are being engineered to ripen more slowly so they will suffer less damage in cross-country shipping, and fast food chains are test-marketing carrot and celery sticks bred by Kraft using advanced tissue culture methods to be crisper and sweeter than the common varieties. Many third world crops, such as coffee, vanilla, cocoa, and some recently discovered sweeteners, are being studied by biotechnologists to see if their flavors can be genetically reproduced in a test tube, thereby cutting the farmers out of the picture entirely.

In animal agriculture, biologists are attempting to change the genetic structure of livestock, so that drugs like BGH [Bovine Growth Hormone, which is designed to increase a cow's milk production] will no longer have to be injected by farmers. Instead, if current experiments pan out, controlled levels of various hormones and drugs will be secreted by the animals themselves. If these animals are successfully patented by drug or chemical companies, which all precedents suggest that they will be, the companies would likely demand a royalty for every offspring of these engineered creatures.

The next step is engineering animals to secrete substances into their milk that are not part of their own usual metabolism, such as drugs for human use. In the fall of 1991, researchers in the U.S. and Europe reported the first successful extraction of human proteins from the milk of genetically engineered sheep, goats and cows. The desired genes are microinjected into artificially inseminated eggs, which are then surgically implanted into surrogate mothers. The success rate has been fairly low so far, and various reproductive and physiological problems have been reported in the test animals. However, the announcements prompted *Bio/technology* magazine to proclaim the age of "barnyard biotechnology," with dairy cows serving as "the ultimate animal bioreactor."

Medical Miracles?

Agricultural applications of biotechnology have aroused opposition in many quarters. It is increasingly clear to people that the development of genetically engineered plants and animals has little to do with feeding the hungry or helping farmers.

Often they are transparent schemes of the large drug and chemical companies—which now also own most of the large seed companies—to increase their control over the world's agricultural production. But biotechnology has managed to sustain a relatively clean public image, largely through its promises of widespread advances in medicine. Advances in medical biotechnology make frequent headlines. We are promised cures for cancer, AIDS, and many heretofore intractable genetic diseases. Should the potential medical benefits allay people's concerns about the hazards of biotechnology?

Some meaningful advances in basic medical research have been obtained through genetic manipulation. So-called recombinant DNA technology has greatly facilitated the laboratory purification of protein hormones, growth factors, and other substances that are found in only the most minuscule quantities in living cells. Scientists can now explore the function of these substances in detail, as test-tube scale bacterial "factories" for their production can make far larger amounts available to researchers than are possible with more conventional biochemical methods. This has aided research in cellular genetics, immunology, and many other fields of study.

However, there are serious limitations. Research in many biomedical fields has become so narrowly focused on these newly accessible molecular factors that other, potentially more clinically valuable, studies are under-funded and often overlooked. It has helped sustain the "magic bullet" approach to medicine, an approach that has been increasingly criticized in recent years, but which remains the key to drug company profits. Combined with the raw manipulative power of gene splicing technology, the uncertainties of research funding have driven many scientists to limit the scope of their work to solving problems through those refinements in genetic technologies that are favored by the pharmaceutical industry. . . .

No Significant Advantages

The newest proposals for genetically engineered drugs are based upon substances that were essentially unknown before the age of biotechnology: the complex protein-based factors that have become a leading focus of biotech research. These substances perform highly specialized biochemical tasks in living cells, and are generally found only in the precise locations where they are needed, often for very short time intervals. They include immunological factors such as interferons and altered antibodies, lymphatic factors such as the interleukins, nerve growth factors, and proteins which are believed to help cancerous tumors alter blood flows deep inside living tissues. A few companies pinned their hopes on a drug called tissue plasminogen activator, or t-PA. This substance has the ability to rapidly dissolve internal blood clots associated with heart attacks, and was promoted as a powerful new cure for a variety of conditions caused by blood clots.

A Dangerous Drug

Bovine Growth Hormone (BGH, also known as BST, for bovine somatotropin) has been a source of continuing controversy in Wisconsin, Vermont and other dairy states for several years. Dairy farmers across the country have opposed this drug, which is designed to increase a cow's milk production. . . . The effects of the drug on the health of dairy cows have been controversial, and the health consequences for people who drink milk with residues of synthetic BGH have barely begun to be investigated. Even the generally staid Consumers Union has raised serious doubts about the safety of milk from cows treated with BGH.

Brian Tokar, *Z Magazine*, February 1992.

T-PA fell far short of the companies' expectations, however. First, the FDA [Food and Drug Administration] delayed approval of the drug due to evidence that it could cause dangerous internal bleeding in some patients. The next obstacle was the price: a single dose of t-PA cost as much as $2200. By the time the drug became widely available at the end of the 1980s, scientists had discovered a number of drugs that proved equally effective at a cost of only a few hundred dollars. Some of these could be extracted from natural sources without resorting to gene splicing methods. One relatively inexpensive drug, commonly extracted from streptococcus bacteria, was found to be as effective as t-PA in several international studies, including one involving over 20,000 heart attack patients in Italy, Belgium and other countries. T-PA demonstrated no significant advantages, either in its overall success rate, or in its ability to guard against complications such as stroke, angina, or secondary heart attacks. . . .

Ignoring Natural Methods

Every few weeks, we read of yet another new drug or new discovery which is to prove once and for all the dramatic medical benefits of the new genetic technologies. A three-year patent dispute between two companies ended in 1991 with the FDA approving a new drug to stimulate red blood cell growth in kidney patients. Another is said to enhance white blood cell production. The biotechnology industry continues to promise that they alone will be able to find a cure for AIDS.

Many products are tested and approved for one specific use, then aggressively marketed "off label" to doctors treating a variety of much more common ailments. Companies anticipate huge future markets for so-called "targeted therapeutics": drugs that are too toxic to be allowed to circulate freely in the body, but which can by various methods be directed to a specific target organ, such as one afflicted with cancer. Meanwhile, a whole host of natural, less invasive methods for treating disease—nutritional therapies, herbal remedies, traditional Oriental medicines, and many others—are found to offer comparable or higher success rates despite their complete rejection by the enforcers of medical orthodoxy. The drug companies continue to have a far greater role in defining therapeutic practices than most patients ever suspect. . . .

The Tryptophan Caper

L-tryptophan is a common amino acid found in all proteins, which has long been an ingredient in natural protein supplements. In the late 1980s, it became accepted as a remedy for sleeplessness, as tryptophan is also a natural biochemical precursor for several neurologically active substances. In 1989, the FDA suddenly pulled L-tryptophan from the market after health officials in New Mexico and other areas began to report that it was associated with a rare, sometimes fatal blood disease called EMS (eosinophilia-myalgia syndrome). Over 1,500 cases were reported, and 27 people died from the disease.

A year later, researchers from the Minnesota Department of Health reported that all of the known EMS cases in that state were associated with particular lots of L-tryptophan manufactured by a Japanese company, Showa Denko K.K. The FDA found that the company had introduced a new genetically engineered strain of bacteria into its manufacturing process in order to increase yields of tryptophan. Batches produced using the new "improved" process were found to be contaminated with unusual dimers of tryptophan, that is, pairs of L-tryptophan molecules joined together by another smaller molecule. The physiological role of these dimers is unknown. . . .

The tryptophan case is especially important in light of the continuing battles at the federal level over the regulation of biotechnology. The only existing regulations were developed by the National Institutes of Health (NIH) in the late 1970s in response to early concerns about the danger of engineered organisms escaping from research laboratories. The NIH guidelines established containment standards for different classes of engineered organisms, and have been relaxed several times since their creation in 1977. With the widespread sale of products of genetic engineering possibly just over the horizon, there are as yet no accepted standards for evaluating the safety of these products.

Since tryptophan was classified as a nutrient rather than a drug, the FDA paid no attention to how it was manufactured until people started dying; the manufacturer has withheld information on their production process asserting that it is all "confidential business information." There is growing pressure from the industry on the FDA and other regulatory bodies to basically overlook new uses of biotechnology in the production of common drugs and other chemicals: if a substance is considered safe when extracted from natural sources, they would have the agency simply assume that genetically engineered versions are safe as well. The case of tryptophan has added a new sense of urgency to efforts by environmental and consumer groups to resist this strategy. . . .

Genetic Screening

Many of the newest applications of biotechnology in medicine do not involve the manufacture of drugs, but seek to commercialize some of the advanced analytical methods that the new genetics has made possible in the laboratory. Scientists have been able to identify genetic markers for a number of diseases by splicing genes from affected people into bacteria and studying the strands of DNA that these patients have uniquely in common. Drug companies are now developing molecular probes to easily screen people for such markers, in some cases before birth.

In a few cases of curable genetic diseases, such screening may allow people to be treated before a condition becomes debilitating. But, although specific genetic "lesions" may be statistically correlated with a particular condition, they do not necessarily reveal the "cause," nor a reliable basis for a cure for the disease. Most human diseases are only partly shaped by genetics; other physiological and environmental factors can be far more significant. Even the genetic component can result from subtle regulatory and developmental problems which would not be detected by simply cloning genes and studying their products. . . .

As biotechnology makes it easier to screen for a wide range of genetic traits, the potential for discrimination based upon such tests also increases. Insurance companies have defended themselves against charges of discrimination in insurance rates on the grounds that such tests provide objective evidence that someone is more likely to contract a particular disease. As researchers begin to find possible genetic correlates for conditions such as heart disease, mental disorders and even alcoholism, the possibilities for unfair discrimination based upon such tests rise dramatically.

There are other serious consequences of expanded genetic screening, especially in the case of diseases for which there is presently no cure. If employers believe they can determine which workers are most likely to be harmed by particular substances in the workplace, there is much less incentive to reduce chemical use or improve safety procedures. Prenatal diagnosis of an incurable disease can create pressure upon parents to choose abortion rather than give birth to an "abnormal" child. Screening for genetic lesions raises new concerns about what is considered "normal" in society, and which conditions are "worthy" of trying to cure. There are also issues of personal privacy at stake, especially as DNA evidence becomes more accepted by judges in criminal trials. All of these concerns are amplified by the tendency of companies to market such tests well before they are shown to be reliable.

An international project is now under way to begin mapping the nearly 100,000 genes found on human chromosomes. The so-called Human Genome Initiative is already the largest single item in the U.S. medical science budget, with current appropriations [as of early 1992] of $100 million. This effort, which represents a major boon for human biotechnology, is already diverting important funds from other areas of medical science, and will encourage scientists to continue playing down environmental, whole-body, and even cellular factors in disease, in favor of those aspects which can be readily studied in terms of molecular-level genetics. . . .

An Unstoppable Technology?

Given the widespread influence and prestige of its leading proponents, plus the billions of dollars of investment capital resting upon the results, many people have come to view biotechnology as unstoppable. The highly technical nature of many of the debates often makes it difficult to sustain public interest. In a time of widespread political disempowerment, many people are willing to trust such matters to the experts, and just hope that the next generation of technology will be able to undo the mess that the previous generation has created. The rates of genetic disease, and of a variety of reproductive and immune disorders, are increasing rapidly, especially among people exposed to high doses of toxic chemicals at work or in their homes. The hazards of biotechnology seem remote compared to the assaults people already face on a daily basis. Never mind that engineering away symptoms may expose people to even greater hazards, while making it more difficult to address root causes or sustain a far less financially lucrative program of real prevention. Perhaps there will be some important benefits in the long run.

It may no longer be possible to put a stop to genetic engineering. Unlike many new technologies, the necessary tools are readily available to most working scientists. Most experiments do not even require any special facilities. Progress using genetic methods has already become the primary gauge of success for many fields of biological and medical research. However, despite lingering uncertainties about the safety of biotech research, the most serious problem is the rush to commercialize new scientific findings well before they are fully understood. Problems like the devastating spread of AIDS are being used to advance the political agenda of deregulation, especially in the pharmaceutical industry, but with implications for food products, pesticides, and industrial chemicals, too. With even academic research agendas increasingly guided by the pursuit of profit, the manipulative power of the new biotechnologies poses a threat to public health, to the quality of the food we eat, and to the biological integrity of life itself.

Just a few years ago, the experts were predicting that biotech products like BGH, genetically engineered plants, anti-frost bacteria, and various exotic medicines would be widely accepted by the early 1990s. The fact that this has not come to pass, that developments in biotechnology are as uncertain and controversial as ever, offers hope that an increasingly educated public will be able to prevent some of the worst consequences of an imperialistic and fundamentally life-denying technology.

"[Prenatal testing's] destructive social consequences may prove to be both far-reaching and long-lived."

Prenatal Genetic Testing Is Harmful

Elizabeth Kristol

One aspect of the growth of genetic engineering is the increased use of prenatal testing to screen for genetic abnormalities. While on the surface this procedure might appear to be purely beneficial, in the following viewpoint Elizabeth Kristol cites many negative social and medical consequences to the continued use of prenatal genetic testing. Kristol is a writer whose articles have appeared in the *New York Times*, the *Washington Post*, the *American Spectator*, *Commentary*, *First Things*, and other national publications.

As you read, consider the following questions:

1. Why is the public health sector motivated to promote prenatal genetic testing, in the author's opinion?
2. Why do women feel pressured to undergo prenatal testing, according to Kristol?
3. In Kristol's view, how will the increased use of prenatal genetic testing affect people's views of the disabled?

Excerpted from Elizabeth Kristol, "Picture Perfect," *First Things*, April 1993. Reprinted by permission of the author.

During the past two decades, prenatal screening for fetal defects has become a standard part of nearly every pregnant woman's medical care. Tests conducted during the first half of pregnancy are designed to detect a wide range of genetic and other disorders, and to give women the option of obtaining abortions if defects are diagnosed. Some people have heralded this development as a breakthrough in the age-old war against disease. Others regard it as more than that: a tool to improve society. Modern birth control methods, the argument goes, brought us quantity control; the addition of prenatal testing offers a system of quality control. For the first time in history, parents are able to customize, albeit in limited ways, the kinds of children they bring into the world.

Prenatal diagnosis may be a routine procedure, but it raises a number of troubling issues. While the women who avail themselves of the tests are usually worried about their children's health, the political, legal, and medical communities have their own reasons for encouraging large-scale screening for fetal defects. Unbeknownst to most prospective parents, moreover, scientists are still debating the safety of the most widely offered screening tests. The ethical issues raised by prenatal screening are even touchier. . . .

Types of Prenatal Testing

The most common form of prenatal testing is ultrasound imaging, which uses sound waves to produce a picture—or "sonogram"—of the fetus. Today, more than 80 percent of all pregnant women in the United States receive a sonogram during their pregnancy. Women deemed at "high risk" for giving birth to a child with chromosomal abnormalities are also offered amniocentesis, a procedure in which a needle, guided by ultrasound, is inserted into the uterus and withdraws a small amount of amniotic fluid for cell analysis. Amniocentesis is usually done between the sixteenth and twentieth weeks of pregnancy. Women may also opt for the somewhat riskier procedure of chorionic villus sampling (CVS), which is usually done between the tenth and twelfth weeks, or earlier on an experimental basis. CVS removes a small amount of chorionic villi (hair-like fringes of the placenta) for analysis, either by using a catheter to pass through the cervix to the womb or by inserting a needle into the abdomen.

Since CVS and amniocentesis are invasive procedures that can harm both the mother and the developing fetus, researchers have long sought a method of testing that cannot endanger mother or child. In the early seventies scientists discovered that high levels of alphafetoprotein (AFP), which is usually leaked from the fetus into the mother's bloodstream in very small quantities, could indicate the presence of neural-tube defects such as anencephaly (incomplete development of the brain) and spina bifida (malformation of the spine), defects that affect 1 to 2 in every 1,000 live births. In 1983 it was discovered that an unusually *low* level of AFP in the mother's bloodstream was a possible indication of Down's syndrome. A simple blood test for AFP is frequently offered to women—regardless of age and known genetic risk factors—between the sixteenth and eighteenth weeks of pregnancy. After ultrasound, it is the second most common form of prenatal testing. . . .

Fear of Malpractice

How is it that perfectly healthy women may find themselves having a series of medical tests, some of which pose distinct risks to themselves or their children? The typical pregnant woman would be disturbed to realize that a good deal of the testing that goes on is motivated by factors that are, at best, tangentially related to her well-being or the health of her child.

The use of AFP tests has a peculiarly nonmedical history. Both ACOG [American College of Obstetricians and Gynecologists] and the American Academy of Pediatrics urged the FDA [Food and Drug Administration] not to approve early release of AFP test kits in the late 1970s. They noted that in order to detect enough cases of open spina bifida and anencephaly the tests would necessarily have a high false-positive rate—about fifty false positives for every true positive. They recommended that the FDA make its release contingent on laboratories' ability to coordinate follow-up tests to weed out false positives, a crucial concern in a test parents may rely on in deciding whether to continue a pregnancy. But when the FDA went ahead and approved the marketing of the kits without these restrictions, ACOG's legal department promptly issued a liability "alert" to its members, urging all obstetricians to offer the procedure to their patients. This, it said, should place the doctor in the "best possible defense position" in the event of a birth defect.

The momentum generated by this single recommendation—inspired by law rather than medicine—was powerful. To offset the inaccuracy of AFP tests, ACOG developed a rigorous protocol for obstetricians. If AFP levels are unusually high, for instance, doctors are urged to repeat the test. If the second test also comes back positive they are to do an ultrasound to determine the reason for the elevated AFP level (such as multiple pregnancy or inaccurate assessment of fetal age). If that is inconclusive, they are to advance to amniocentesis. If that is abnormal, they are to perform a high-resolution ultrasound. With each subsequent test, there is an increased chance that any number of anomalies, slight or severe, may be detected. Thus, a patient who follows her doctor's suggestion to undergo testing for neural-tube defects might find herself, a few weeks down the line, being counseled to contemplate an abortion for a variety of lesser disorders for which she had no original intention of seeking testing.

Reducing the Number of Deformities

Like the medical community, the public health sector has its own reasons for promoting widespread prenatal screening. The U.S. Department of Health and Human Services has announced a goal of screening at least 90 percent of the U.S. population "for fetal abnormalities," an objective that "will be measured by tracking use of maternal serum alphafetoprotein screening tests." The HHS report that explains this goal states that "current ACOG standards recommend that MSAFP screening be offered to all patients"—without noting that this was a legal, not medical, recommendation. Likewise, the California Department of Health, as part of its ambitious statewide screening program, requires everyone who offers prenatal care to inform pregnant patients of the AFP test in an effort to detect greater numbers of potential birth defects. The fact is that governments on both the state and national level have considerable interest in being able to point to reductions in disease. And morbidity and mortality rates are key expressions of a region's standard of living.

When most people hear of "reducing illness," they usually think of providing greater access to health care or developing new treatments for disease. Public health experts, however, frequently boast of reducing illness by means of prenatal diagnosis and abortion. . . .

Policymakers and medical experts are under pressure not only to achieve noticeable improvements in health but also to reduce soaring health care costs. Widespread prenatal screening followed by abortion for fetal defects would accomplish both of these objectives. The motivation to reduce costs also helps explain the long-standing emphasis on preventing the birth of children with Down's syndrome, a disorder that is more financially costly to society—accounting for about 15 percent of the institutionalized mentally retarded population—than it is personally costly to its victims. (There are certainly other disorders and diseases that cause greater pain and discomfort.)

In the 1950s and 1960s, when studies seemed to indicate that more than half the children with Down's syndrome were born to mothers over the age of thirty-five, women over thirty-five were urged to have amniocentesis. When two decades of screening and abortion of Down's fetuses in this age group failed to have a significant impact on the national Down's syndrome population, new studies were undertaken. These revealed that only about 20 percent of Down's children are born to women over thirty-five, and that in many cases (nearly a quarter, according to one study) the father may be the source of the extra chromosome that causes the disorder. By itself, then, amniocentesis of women over thirty-five would not do the trick. The discovery that Down's syndrome could also be detected by the AFP blood test, which is safe enough to be given to all pregnant women, was therefore regarded as a major breakthrough.

There has been no shortage of arguments to eliminate the ill or disabled before they become a financial burden to society. In a survey of British obstetricians in the late 1970s, researcher Wendy Farrant discovered that two-thirds of the respondents rated "savings in costs to society of caring for people with disabilities" as an important benefit of a national screening program for neural-tube defects; 13 percent agreed that "the state should not be expected to pay for the specialized care of a child with a severe handicap in cases where the parents had declined the offer of prenatal diagnosis of the handicap." More recently, the British Royal College of Physicians recommended a nationwide program of prenatal screening on the grounds that cost-benefit analysis showed that "it is cheaper to screen and counsel the whole population than it is to treat affected children who would otherwise be born to unprepared couples."

The New Discrimination

Increasingly in the 1990s we are going to see a virulent new form of prejudice and discrimination emerge, based on one's genetic type. This is going to be a much more dangerous form of discrimination than skin color, than religion, than ethnicity.

Jeremy Rifkin, in Joseph Levin and David Suzuki, *The Secret of Life*, 1993.

Medical cost-benefit analyses are startlingly cold-blooded. Studies feature graphs comparing the costs to society of a disabled child with the expense of testing and abortion. Articles debate the appropriate discount rate that should be used in calculating the lifetime costs to the state of caring for a disabled individual. One recent study, which noted the growing cost of providing services for mentally handicapped young adults, lamented the increase in the number of patients with Down's syndrome—an increase the authors attributed to medical advances that have allowed those with Down's to live longer and healthier lives. Debate has surfaced within the cystic fibrosis community over whether advances in the comfort and lifespan of individuals with CF outweigh earlier arguments favoring abortion of fetuses diagnosed with the disorder.

Crucial to all the discussions, reports, and studies supporting prenatal testing is the assumption that women will have abortions if fetal defects are detected. The hard truth is that there are still very few conditions that can be treated in utero. Hospitals will occasionally do fetal blood transfusions or perform surgery for urinary tract obstruction, and drug therapy is useful for treating some metabolic diseases. Experimental research in the area of gene therapy, the replacement or correction of a defective gene in the fetus, would open up the possibility of new forms of prenatal treatment. For the foreseeable future, however, the chief purpose of prenatal diagnosis is to give parents the opportunity to abort a fetus diagnosed with a disorder. It is telling that research in the area of prenatal diagnosis is overwhelmingly concentrated on finding ways to diagnose conditions in the first few months of pregnancy, when abortion is a simpler and safer procedure, even though information about the fetus is much richer later on.

Yet the "A" word is almost never mentioned in the screening literature. When allusion to the subject is unavoidable, it is glossed over with an extraordinary amount of euphemism. This is the case even in medical journals, where doctors are addressing one another rather than pregnant patients. Physicians refer to "screening and its sequelae." Pregnancies are "terminated," "selectively terminated," or, most bewildering, "interrupted." Parents who receive news of a fetal disorder are urged to "choose a reproductive option," "decide the disposition of their pregnancy," or, simply, "intervene." In discussing abortion procedures, physicians refer to "permanent asystole" or "mechanical disruption of the fetus" rather than fetal death. The word "amniocentesis" often serves as a stand-in for testing-plus-abortion; one genetics textbook states, "If all mothers of thirty-five years and over had amniocentesis then this would reduce the incidence of chromosomal disease by 30 percent." Many British physicians take recourse in acronyms, referring simply to "TOP"—termination of pregnancy. . . .

The Pressures to Test

While many pregnant women welcome the choices prenatal testing has given them, others are ambivalent, have misgivings, or have simply not given the matter much thought. Yet the pressures to be tested are powerful. The most obvious pressure comes from the context in which tests are offered. Studies show that even women who have reservations about screening find it difficult to decline tests when their obstetricians suggest them. In one survey, about a third of the women who had already agreed to be tested "had wondered if it was right to perform a kind of quality control of the fetus."

In the doctor's office and in the many popular books available on pregnancy and childbirth, there is an assumption that reasonable and enlightened women will naturally want to make use of new screening technologies. The 1983 President's Commission on genetic screening is typical in describing prenatal testing and carrier screening (the testing of couples before conception to determine whether they carry a genetic defect) as enhancing a woman's choices. "Genetic screening and counseling are medical procedures that may be chosen by an individual who desires information as an aid in making personal medical and reproductive choices," it says. "Professionals should generally promote and protect patient choices to undergo genetic screening and counseling. . . ."

Politicians and pollsters have long known that the words "information" and "choice" are powerful ones for Americans—especially for women. Barbara Katz Rothman, a sociologist at Baruch College in New York, has observed that we are raised to welcome all offers of both: "If there is information to be had, and decisions to be made, the value lies in actively seeking the information and consciously making the decision. To do otherwise is to 'let things happen to you,' not to 'take control of your life.'" Women who reject screening are regarded as "turning *away* from the value of choice, and even more profoundly, turning away from the value of information."

Increased Anxiety over Pregnancy

Doctors, however, don't have to live with the anxiety generated by testing and the gathering of information; patients do. Yet physicians and women's health advocates repeatedly insist that the best reason for women to undergo prenatal screening is for "the reassurance it almost always brings." This is a strange assertion. Certainly, worrying is a natural part of any pregnancy: Can my body do all the things necessary to carry the baby to term? Will the baby be healthy? Will I be a good parent? Such free-floating concerns have always plagued women. But in the past few decades, the normal anxieties of pregnancy have been inflamed by a highly specific set of specters—specters prompted less by genuine health threats than by the promotion of certain tests.

Because there is a test for Down's syndrome, for example, women over the age of thirty have been bombarded with articles about the risks of having a child with Down's; many women can chant the statistics for each age category. To look at this situation from afar, one would assume that women today are at increased risk of giving birth to a child with Down's, or that Down's syndrome accounts for a majority of birth defects or, at least, a majority of cases of mental retardation. In fact, Down's syndrome accounts for only a fraction of all birth defects (including mild retardation) and only a quarter of the cases of serious retardation, which can be caused by a number of unpredictable genetic factors as well as trauma during the birth process. Similarly, the other chromosomal abnormalities, fetal infections, neural-tube defects, and blood and metabolic disorders that can currently be diagnosed before birth do not begin to exhaust the universe of possible defects.

Women have been trained to concentrate their anxieties on Down's syndrome for the simple reason that they are offered tests for it. But they are offered tests for Down's, not because the risk is personally high for them, but because the public health sector has a powerful interest in reducing the number of citizens who may end up requiring government support. Major research efforts have therefore been concentrated on screening for Down's, one of the few forms of mental retardation whose cause is known.

Displaced anxiety can lead to artificial peace of mind. In the current climate of testing it is all too easy for prospective parents to forget that illness can befall a baby at any time during pregnancy and delivery, or after birth, and that the majority of birth defects are undetectable and unpreventable. Yet, as obstetricians will be the first to admit, many women who receive a negative result on a prenatal test seem to feel that they are in the clear. This false sense of security can make an undiagnosed birth defect or subsequent childhood illness all the more difficult to handle. . . .

Negative Views of Disability

As prenatal screening becomes increasingly routine, disability ceases to be viewed as a random misfortune. But even if a woman had all the reproductive choices in the world—whether to conceive, whether to undergo diagnostic testing, whether to treat the fetus, or whether to abort for a particular condition— she still would not be guaranteed a healthy child. When children are born with disabilities or suffer injuries in childhood, will parents steeped in a culture of screening regard them with resentment? The effect of this culture, Barbara Rothman has pointed out, is that conditionality, rather than acceptance, is built into parental love from the start. Screening for defects is a way of saying: "These are my standards. If you meet these standards of acceptability, then you are mine and I will love and accept you totally. After you pass this test." Pediatrics expert Jeffrey Botkin agrees that screening may have a destructive effect on the parent-child relationship, noting that testing raises parents' expectations of their children, rather than encouraging parents to recognize the uniqueness of each child.

Disability advocates and feminists interested in the social impact of reproductive policies have criticized society's growing role in developing and enforcing quality-of-life standards. Even some feminists who are resolutely pro-choice have trouble with abortion for defect. As Harvard's Ruth Hubbard has explained, "It is one thing to abort when we don't want to be pregnant and quite another to want a baby, but to decide to abort this particular fetus we are carrying in hopes of coming up with a 'better' one next time." Disability groups and feminist supporters fear that when physicians encourage the abortion of fetuses with diseases or disabilities, they are fostering intolerance of the less-than-perfect people who are already born. Anecdotal evidence gives cause for concern: in one study of seventy-three parents-to-be undergoing prenatal screening, 30 percent said they thought screening might encourage negative attitudes toward the disabled; half thought that mothers of disabled children would be blamed for their failure to undergo screening or have abortions.

University of Wales geneticist Angus Clarke has remarked on the poisonous effect of the double standard that governs prenatal screening. Physicians and policymakers, he notes, assume that abortion for sex selection is "tantamount to a declaration that females are of much less social value than are males. Society is not willing to make such a statement, which would have profound implications for how women are viewed in society, and also for how women view themselves." Yet there are no restrictions on the patient's autonomy to abort for any disability whatsoever. This, Clarke says, indicates the "low value that our society places upon those with genetic disorders and handicaps. We draw some moral lines for social but none for genetic termination of pregnancy.". . .

Uncharted Dilemmas

As screening becomes increasingly widespread and sophisticated, physicians, policymakers, and the courts will be forced to make judgments about what kind of life is worth living and what kinds of disabilities are too costly to society. Already, parents who undergo prenatal testing are finding that answering life-and-death questions is more difficult than they had imagined. How "normal" does a baby have to be to continue the pregnancy? Which is worse, a severe physical or slight mental handicap? Should one abort if there is a 30 percent chance that a genetic disease will be transmitted? Is it worth giving birth to a child who will die at the age of forty? Thirty? Twenty?

Prenatal testing has the potential to raise countless uncharted dilemmas. If parents who choose to abort in the case of a detected defect already have children, how do they explain the sudden disappearance of the pregnancy? Do they tell the children it was a miscarriage, or do they try to explain that the pregnancy was ended because the baby had an illness? Other, more peculiar, situations present themselves when mild or ambiguous disorders are diagnosed and parents choose not to abort. In the case of conditions that may affect growth, sexual development, or level of aggression, Rothman has noted, parents might find themselves locked into a certain perception of their children, always on the lookout for signs of abnormality. Perfectly normal childhood behavior will be scrutinized for manifestations of certain diseases. There is no way to know how this atmosphere might affect a child's development and sense of self. As the ability to detect a wider range of nonfatal genetic conditions becomes possible, these sorts of challenges may become increasingly common.

Rothman has also described the daunting problem posed by the detection of late-onset disorders, such as Huntington's disease, that do not manifest themselves until adulthood. If parents know the awful secret that their child probably will not live past a certain age, how will this knowledge affect their relationship with the child? Will they find themselves keeping an emotional distance to protect themselves from future pain? Will they, consciously or unconsciously, skimp on ways they invest in their child—whether in education or in encouragement of talents, hobbies, and other skills?

The decisions raised by prenatal testing are the stuff of moral philosophy. But they put real-life parents in inhumane situations. Moreover, they coarsen our very notions of what is involved in being a parent and what it means to be a responsible member of society. Through the gradual introduction of new forms of technology and testing, the medical establishment and the public health sector have been developing subtle quality-of-life standards and not-so-subtle ways of discouraging the birth of those who do not measure up. Debate on the issues raised by screening, when it does take place, has been confined to a small circle of professional ethicists, legal scholars, and feminists interested in reproductive policy.

Testing for birth defects, meanwhile, has crept into the life of nearly every woman of childbearing age, whether she avails herself of it or not. It is not too strong to say that childbearing has, in a profound sense, been transformed. This transformation is not the province of one interest group or another: it is not exclusively a medical issue, a legal issue, an economic issue, or a women's issue. Like many revolutions in medicine and technology, prenatal testing took on a life of its own before its implications could be fully assessed. Like too many revolutions, its destructive social consequences may prove to be both far-reaching and long-lived.

> "Prenatal screening programs will significantly
> reduce disease and infant morbidity, thereby
> contributing to the overall state of national health."

Prenatal Genetic Testing Should Be Universally Available

David T. Morris

Prenatal genetic testing is an effective way to reduce health care costs and increase women's reproductive choices, David T. Morris writes in the following viewpoint. He asserts that prenatal testing should be made available to all women as a way to improve America's health and to give women increased information concerning their pregnancies. Morris, an attorney, is a graduate of Boston University School of Law.

As you read, consider the following questions:

1. What are the three main types of prenatal genetic testing, and what are their advantages and disadvantages, according to Morris?
2. How does prenatal genetic testing "foster reproductive choice and individual autonomy," in the author's opinion?
3. What reasons does Morris give for the proposal to provide access to prenatal testing and abortion to all American women?

Abridged from David T. Morris, "Cost Containment and Reproductive Autonomy: Prenatal Genetic Screening and the American Health Security Act of 1993," *American Journal of Law and Medicine*, vol. 20, no. 3 (1994), pp. 295-316. Reprinted with permission of the American Society of Law, Medicine, and Ethics.

Prenatal genetic diagnosis represents one of the most important recent advances in the field of clinical genetics. Each year in the United States, a significant number of children are born with some type of hereditary genetic defect that can be diagnosed *in utero* [in the womb]. Of the over 4000 genetic traits which have been distinguished to date, more than 300 are identifiable via prenatal genetic testing. With increasing frequency, various diagnostic procedures are being used to evaluate the probability that a fetus will be born with a serious physical or mental handicap caused by a genetic abnormality. These procedures identify disabling or potentially fatal genetic disorders, including cystic fibrosis, trisomy 21 (Down syndrome) [the most common cause of moderate to severe mental retardation], ß-Thalassemia [an inherited lethal disease characterized by deficient production of hemoglobin, which leads to impaired development, infertility, and reduced longevity], and neural tube defects (NTDs) [which include anencephaly, in which the affected child is born without all or part of the brain, and spina bifida, which is marked by a defect in the development of the spinal column which leaves part of the spinal canal exposed].

Prenatal diagnosis may be achieved by utilizing one or more techniques from three basic methods of testing. The most commonly performed means, amniocentesis, is a procedure in which a small sample of the fluid surrounding the developing fetus is extracted from the amniotic sac. The fluid, which contains cells from the fetus, may be analyzed after removal to determine fetal sex as well as to locate specific genetic abnormalities. Although utilization of amniocentesis at earlier stages of gestation has been investigated, the procedure is usually not performed until the sixteenth week of pregnancy. Test results are generally available within two weeks, though in some instances the wait may be as long as four weeks.

Other Tests

Amniocentesis is rapidly being replaced by another form of fetal tissue extraction, chorionic villi sampling (CVS). CVS involves transcervical or transabdominal insertion of a catheter to an area of placental development, where a small sample of placental tissue is removed by aspiration. The advantage of CVS is that it can be performed as early as nine weeks into the pregnancy. Moreover, because test results are usually available within two or three days, CVS allows first-trimester diagnosis of any identifiable chromosomal abnormalities or genetic diseases.

The third method of prenatal diagnosis is maternal serum alpha-fetoprotein sampling. Unlike amniocentesis and CVS, which are physically invasive procedures, alpha-fetoprotein sampling merely requires the woman to undergo a blood test.

High or low concentrations of alpha-fetoprotein in the mother's blood may indicate a risk of fetal genetic abnormality. The usefulness of alpha-fetoprotein sampling is currently limited to the diagnosis of neural tube defects, although recent studies have indicated that low concentrations may be associated with increased risk for Down syndrome. Moreover, significant numbers of false positives have been known to occur, thereby necessitating additional diagnostic tests.

Targeting Tay-Sachs

Properly designed and implemented genetic screening procedures can have positive results, as evidenced by a program aimed at Tay-Sachs disease. As many as one in twenty-five Jews of Eastern European descent is a carrier of this disorder, characterized by nervous system degeneration, uncontrollable convulsions, total loss of sensory input, and death, usually by age four. . . . The screening program succeeded in lowering the incidence of the disease by 90 percent between 1970 and 1992.

Joseph Levine and David Suzuki, *The Secret of Life*, 1993.

The impact of genetic disease "in terms of social, emotional, and financial costs is usually greater than that of other [nongenetic] illness," according to Arno G. Motulsky. Nevertheless, and despite rapid advances in genetic technologies, "genetics simply has not become an integral part of the public health scene in the United States," according to F. John Meaney and Susan P. Chang. This is in large part evidenced by the fact that the federal budget for genetic service programs has remained unchanged since 1981. In fact, in 1981, following the repeal of the Genetic Diseases Act of 1976 [which was designed to deliver genetic disease information and to provide education, testing, counseling services, and medical referral for all persons who were suspected of having or transmitting a genetic disorder], federal funding for genetic services was reduced approximately forty-seven percent. Such policy choices have effectively limited access to prenatal screening services to individuals who are primarily white, middle class, and well-educated. . . .

Reasons for Prenatal Testing

Prenatal genetic screening services are often justified on the grounds that they foster reproductive choice and individual autonomy. Prenatal genetic screening encourages effective choice by augmenting the number of reproductive options available to individuals who are bearing or who have considered bearing children. By increasing the amount of information available during pregnancy, prenatal diagnosis provides individuals the opportunity to make more informed decisions regarding childbirth. A diagnosis may be made early enough to allow a woman who is carrying an affected fetus the opportunity to terminate the pregnancy safely within the first trimester. Prenatal diagnosis also affords the parent of an affected fetus who does not wish to abort the chance to prepare for the birth of a child with very special needs. This may include arranging to meet the financial demands that attending to the child's disorder will undoubtedly place on the parent(s). It may also give the parent and other family members time to familiarize themselves with the nature of the disease and the special emotional and medical needs associated with caring for an afflicted child.

In a sense, prenatal diagnosis enables parents to fulfill their personal expectations with respect to raising a healthy family. This is especially true for parents who have a known risk of bearing a child with an inherited disorder and who might otherwise decide to forego having children. The average couple with a known genetic risk "may convert a statement of risk . . . into a binary statement (either it will or will not happen). They then may visualize the worst outcome," according to Peter T. Rowley. However, prenatal diagnosis "converts a probability statement . . . into a statement of fact that the fetus has the disease, or it does not," according to Charles R. Scriver. To the extent that a diagnosis precludes an adverse outcome, the opportunities provided by prenatal screening may thus be viewed as "pro-life."

Cost Containment

Cost reduction . . . may also justify furnishing prenatal genetic screening services to a greater segment of the population. The costs associated with raising and caring for a child affected with an inherited disease can be very high. As a result, according to Harvey R. Colten, they "constitute a major burden for patients and their families and have an effect on the health care delivery system that is disproportionate to the absolute numbers of patients" affected with a serious hereditary disease. For example, over forty percent of the more than 15,000 persons affected with cystic fibrosis require expensive hospitalization for at least a week or more each year. Medical costs for children affected with spina bifida may reach well over $80,000 for the first year of life alone. The cost of caring for a single child with Down syndrome is estimated to be in excess of $500,000.

Compared to these costs, the expense of undergoing prenatal screening is very small. For example, a woman undergoing an amniocentesis can expect to pay somewhere in the area of $1000 for the test. In fact, research has demonstrated that programs incorporating prenatal diagnosis have been highly cost-effective in reducing the financial burdens accompanying the birth of a child born with a serious hereditary disease. These studies indicate that the financial outlays for administering screening programs and performing abortions when a fetus is found to be affected are less than the amount which would be spent caring for affected children. As such, they strongly support a utilitarian approach to incorporating prenatal genetic screening services into a package of basic health care services.

Nevertheless, screening programs identified solely by their ability to reduce health care expenditures strongly tend to undermine women's rights to self-determination and reproductive autonomy. Cost-benefit justifications have frequently been acknowledged by health economists as a useful means of allocating health care resources. However, while " [c]ost-benefit analysis can make a useful contribution to allocational decision-making . . . it does not provide a means of avoiding difficult [personal and] ethical judgments," according to the 1983 President's Commission for the Study of Ethical Problems in Medicine and Biomedical and Behavioral Research, Screening and Counseling for Genetic Conditions. Cost-benefit justifications for prenatal screening programs are suspect in that they fail to quantify what the President's Commission calls the personal and "psychological 'costs' and 'benefits' to screenees, their families, and society." They effectively supplant uniquely individual decision-making criteria with the economic goals and considerations of others. In so doing, programs whose purposes are grounded in such justifications undervalue the intensely personal and emotional character of reproductive decision making. . . .

Preventive Care

The potential to reduce or eliminate the incidence of severely disabling or potentially fatal disorders is perhaps the most compelling justification for the use of prenatal genetic technologies. According to expert Judith G. Hall, genetic disorders "are responsible for a major portion of morbidity and mortality" among neonates. Clinical geneticists and researchers in particular believe that as a tool for prevention, effective prenatal screening programs will significantly reduce disease and infant morbidity, thereby contributing to the overall state of national health. Most people would, without question, embrace a program designed and able to prevent or eliminate disease. Few, if any, public health programs could be politically or economically justified were it not for the fact that their activities served such ends. Assuming that prenatal screening can, in fact, reduce the incidence of genetic disease, there are still several reasons to question the goals of a prenatal screening program solely premised on such a "public health model."

First, in the context of prenatal screening, prevention necessarily assumes termination of the pregnancy. Medical technology has yet to advance to the stage where *in utero* fetal gene therapy may be undertaken to effectively correct or diminish the effects of a genetic abnormality. Accordingly, such a model completely ignores the fact that for many women abortion is not an option. . . .

Ambiguities

Prevention also appears to be an unsatisfactory rationale for prenatal screening programs when it is considered that a fair amount of ambiguity is present in the diagnosis of an unborn child. Many diseases identifiable through prenatal screening vary in the severity of their effects. A fetus diagnosed with Cystic Fibrosis, for instance, might die as an infant, be severely disabled with chronic obstructive pulmonary disease, or live an athletic lifestyle and survive into middle age. The power of prenatal screening is limited to *diagnosis*, and does not afford a useful *prognosis* upon which to evaluate the child's potential quality of life. Consequently, prenatal screening programs which target prevention or elimination of disease as a goal foreclose consideration of whether an affected child may lead a rewarding or valuable life.

On a social level, this same result is manifested in society's conception of disease and its willingness to accommodate those who are disabled by it. Disease is as much a social construct as it is a physical reality. "[N]ot everyone will arrive at the same conclusion about the classification of human genetic differences because different views may be held about what justifies classifying a trait or characteristic as a disease, an abnormality, or a healthy state," according to Arthur L. Caplan. Acceptance of prenatal screening solely as a means of eliminating disease may indicate society's hostility towards accepting disabled persons as valuable members of society. This in turn stigmatizes persons living with genetic disease. Conceivably, "[a]s our society develops ever more refined prenatal diagnosis and medical technology, more and more characteristics may be considered unacceptable deviations and disabilities," according to Adrienne Asch. The significance of these concerns becomes apparent once it is recognized that it may be impossible to wholly eradicate the existence of genetically-linked disease.

Lastly, a prevention-oriented screening program replaces the individual's appreciation for the value which may inhere in raising a disabled child with a social mandate in favor of discrimination. It is questionable whether the value society places on individual choice can be preserved where public health goals stress a reduction in the number of children born with mental and physical impairments. While decisions regarding whether or not to abort an affected fetus may, as Asch puts it, "depend on our notions of what sorts of lives are worthwhile," they are as much personal decisions as they are social. Thus, at least in the context of prenatal diagnostic programs, goals explicitly premised on the prevention or reduction of disease must be framed within a setting which accommodates such personal decisions if consumer choice and autonomy are to remain legitimate concerns of any health reform plan. . . .

Screening and Health Care Reform

Prenatal genetic screening should be a part of any national health care reform plan's guaranteed benefits package. . . . Its value to reproductive autonomy cannot be understated. Providing such services as part of the minimum package of benefits ensures that all segments of the population have equal opportunities to exercise reproductive autonomy. Recourse to such services should not, however, be made mandatory. Any program of compelled testing would not only undermine consumer choice, a central tenet of reform, but would also have serious adverse effects on individual rights to self-determination and informed consent.

Abortion must also be included within the guaranteed benefits package. It is consistent with the overall objectives of the national health care plan. It is also necessary for the effective provision of prenatal genetic screening programs. Significant public resources have already been devoted to the technology making such diagnostic tests possible. To eliminate the availability of abortion to those who are to have access to such programs would effectively render such technologies useless. Without the ability to choose termination, reproductive autonomy has no meaning. Equal access to health care should mean equal access to the means of exercising personal, responsible reproductive decisions.

Special care, however, should be taken with respect to defining goals and priorities in the context of genetic screening services. Emphasis on priorities such as cost control and prevention create social and ethical tensions. The provision of screening services will assist in facilitating the goals of prevention and cost control. However, the ultimate decision as to how to make use of such resources should be left to the individual. In a rapidly changing health care environment, the pressures of limited resources are likely to point toward services that reduce expenditures rather than those which inform and facilitate individual choice. Dialogue between policy makers and consumers is necessary to determine what indications will justify prenatal diagnosis. . . .

Universal access to prenatal genetic screening services may serve to facilitate a number of important objectives. Neonatal screening expands reproductive options. It may also serve to advance public health efforts as well as to provide a cost-effective means of dealing with the financial expense of raising an affected child. Nevertheless, any national health care plan designed to emphasize preventive medicine and cost control must treat genetic screening services with a special degree of care. In the rush to contain health care costs, the rights and needs of individuals should not be overlooked. Attention must be directed towards balancing the projected goals of any such plan with an individual's right to reproductive autonomy and self-determination.

Periodical Bibliography

The following articles have been selected to supplement the diverse views presented in this chapter.

Nicholas Agar — "Designing Babies: Morally Permissible Ways to Modify the Human Genome," *Bioethics*, vol. 9, no. 1, 1995. Available from 108 Cowley Rd., Oxford OX4 1JF, England.

George J. Annas — "Mapping the Human Genome and the Meaning of Monster Mythology," *Emory Law Journal*, Summer 1990.

John J. Conley — "Narcissus Cloned," *America*, February 12, 1994.

Jack Doyle — "The End of Nature?" *Christian Social Action*, January 1991. Available from 100 Maryland Ave. NE, Washington, DC 20002.

Elaine Draper — "Genetic Secrets: Social Issues of Medical Screening in a Genetic Age," *Hastings Center Report*, July/August 1992. Available from 255 Elm Rd., Briarcliff Manor, NY 10510.

Gail Dutton — "Genetic Engineering's Brave New World," *The World & I*, August 1991. Available from 2800 New York Ave. NE, Washington, DC 20002.

Jeff Elliott — "Ethicists Inherit Gene Dilemma," *Insight*, March 27, 1995. Available from 3600 New York Ave. NE, Washington, DC 20002.

Stephen S. Hall — "James Watson and the Search for Biology's 'Holy Grail,'" *Smithsonian*, February 1990.

Mark O. Hatfield — "Stealing God's Stuff," *Christianity Today*, January 10, 1994.

Erik Lindala — "Renewed Debate Surfaces Around Human Genome Project," *Alternatives*, September/October 1994.

Ingmar Persson — "Genetic Therapy, Identity, and the Person-Regarding Reasons," *Bioethics*, vol. 9, no. 1, 1995.

John A. Robertson — "The Question of Human Cloning," *Hastings Center Report*, March/April 1994.

James D. Watson — "The Human Genome Project: Past, Present, and Future," *Science*, April 6, 1990.

How Does Genetic Engineering Affect Agriculture?

Chapter Preface

When most Americans go to the grocery store to buy tomatoes, they find hard, pale-red objects that are tasteless and barely worth buying. This is because most tomatoes must be picked before they are ripe to prevent them from rotting during shipping. While vine ripening is a key to creating a deep-red, juicy, flavorful tomato, until recently it has not been possible to ship ripe tomatoes.

Enter the Flavr Savr Tomato, the first genetically engineered food product to be approved for commercial sale. Biotechnologists have inserted a gene into the tomato that delays the rotting process so that the tomatoes can ripen on the vine and will not rot during shipping.

The Flavr Savr exists because of genetic engineering. Such technology might also someday produce leaner pork and pest-resistant crops, among other agricultural innovations. At first glance it would seem that everyone would welcome these advances in agriculture. As science writer Gail Dutton states, "Transgenic plants, animals, and microbes are just beginning to be released for field tests, and their success offers many benefits."

But many scientists and agricultural experts worry that there has been inadequate research on how genetically engineered plants and animals might affect consumers and the environment. They argue that altering an organism's genetic makeup could affect how it interacts with other organisms, including both humans that consume it and other plants and animals in its habitat. Writer Dick Russell concludes that "we would be wise to apply the brakes to the transgenic food steamroller."

There is much debate about the implications of altering an organism's genes to improve it for human consumption. The contributors to the following chapter present their views concerning the benefits and dangers of applying genetic engineering to agriculture.

"The green-gene technology that created a tastier tomato will also benefit the environment and help feed the 10 billion mouths that will be here within half a lifetime."

Genetic Engineering Improves Agriculture

John Dyson

Genetic engineering can help farmers produce heartier and tastier fruits and vegetables, John Dyson writes in the following viewpoint. By altering plant genes, scientists can also create bug- and herbicide-resistant cotton and plants that produce biodegradable plastics and human proteins for medical treatments. Dyson disagrees with those who fear the consequences of biotechnology, arguing that regulations are in place to prevent the abuse of this new technology. Dyson is a writer and contributor to *Reader's Digest*.

As you read, consider the following questions:

1. What is the difference between genetic engineering and traditional crossbreeding, according to Dyson?
2. How were scientists able to create a better tomato, according to the author?
3. How might transgenic plants help improve human health, in Dyson's opinion?

In January 1992, when we were still eating those pink vinyl rocks that pass for tomatoes, a California scientist handed me a tomato as red as a fire truck and as juicy as any backyard beauty in August. One bite and I was transported to summer's ecstasy.

This delicacy, which you will find at your grocer's, is the first blockbuster product of "plant biotechnology," a new science that's expected to revolutionize agriculture. Besides tastier tomatoes, farmers may be growing crops year-round that defy drought, make their own fertilizer, and manufacture medicinal drugs. It sounds like science fiction, but it's not.

In 1980, biotechnology hardly existed, though the groundwork was laid in the 1970s when scientists learned to cut genes and move them from one organism to another. Soon, they were making bacteria churn out valuable proteins and learning to correct the gene defects that cause human disease.

Today, biotechnology is growing faster than ever. Research suggests that a fifth of all food may be produced with biotech by the year 2000—and not a moment too soon. To feed the world's burgeoning population, says Richard D. Godown, president of the Industrial Biotechnology Association, "biotechnology is our best hope."

A "Green-Gene" Revolution

There is nothing fundamentally new about bioengineering. Every seed catalogue reveals the way growers have tinkered with plants over centuries, crossbreeding them to improve yield and variety.

What is new is the speed at which crossing can now be done. It took centuries for man to develop corn from a wild grass. Today, a new variety of softer or sweeter corn could be designed and grown in two or three years.

Tastier tomatoes herald only the beginning of a tremendous "green-gene" revolution. Our fast-developing ability to program plants for specific purposes—to slow the rotting process, for instance—is already a pathway to incredible opportunities.

Supermarket tomatoes have to be tough and tasteless because nature is too quick. The acids, sugars and aromas that suffuse a tomato with tangy flavor are the last things it gets. From that moment, all the fruit wants to do is soften and burst so its seeds will scatter.

To beat the softening, growers pick tomatoes green. But even with refrigeration, which kills any flavor the tomatoes might have gained, a third are spoiled by rot.

The softening is caused, in part, by an enzyme known as PG, which dissolves the glue holding the tomato's cells together. Like a brick building when its mortar erodes, the tomato collapses. So the challenge to put fresh taste back into the $5-

billion-a-year fresh-tomato market has been to block or slow down the PG enzyme.

Tomatoes, like humans, have about 100,000 genes—chemical commands packed into every cell. The problem facing microbiologist Bill Hiatt of California's Calgene company was to switch off the single gene that sets the PG enzyme to work.

You can't snip out a gene that you want to turn off, says Hiatt, but you can deactivate it by inserting a copy of the gene made in reverse. "Think of the gene as an enormously long train with the locomotive as its on-switch and the caboose its off-switch," he explains. "We reversed the order of the freight cars, then hitched them up again."

The Potential of Plants

We grow plants because they make useful products—food, for example. They could be used as much more versatile biological factories. There is already the prospect of using potatoes and tobacco plants to make antibodies and other human blood proteins.

Steve Jones, *The Language of Genes*, 1993.

From a few cells growing on a dish under lab lights, the transformed plants were cultivated and then grown in a greenhouse. Ten months later, the first ripe tomatoes were picked and put to the test.

"The PG enzyme had been blacked out by its own mirror image," Hiatt says. "By switching it off, we bought the tomato an extra week on the vine. Now, we no longer have to pick it green."

It is only a matter of time before bioengineers bring tastier varieties of other soft fruits to market, including melons, peaches, strawberries and raspberries. "Soon every fruit and vegetable will not only taste better," says Donald Helinski of the University of California at San Diego, "but grow better, travel farther and last longer."

Flounder Antifreeze

Nature's chemical language is the same in all organisms—plants, animals, insects, microbes, humans—so all genes are theoretically interchangeable. A tomato given the gene of a fungal-fighting microbe can fend off the gray mold called botrytis. A potato given the gene of the unappealing petunia makes the Colorado beetle reject it.

Potatoes are the largest single vegetable crop in the world, but vast quantities are spoiled by decay. Now, Bill Belknap of the

Department of Agriculture's research station in Albany, California, is searching for a way to make them rot-resistant, using genes from chicken embryos and insect immune systems.

Then there's the strawberry that might "marry" a flounder. Strawberries are often wiped out by frosts. Learning that the Arctic flounder makes an antifreeze to protect itself against winter chills, bioengineers at DNA Plant Technology Corp. in Oakland, California, plan to inject the antifreeze gene into a strawberry, so the plant would be able to make the antifreeze for itself. Presto! Strawberries that survive frosts and don't go mushy when you take them out of the freezer. . . .

So far, only a few plant genes have been isolated, but the range of possibilities is boundless. Nearly a fifth of all the coffee drunk in the United States is decaffeinated, mostly by chemicals. But in the future it may be done naturally, on the tree. Grapefruit may be sweetened on the branch and popcorn given a buttery flavor on the cob, so both could be enjoyed without the extra calories added by oil or sugar.

Special Products

As more of these novelties are developed, plants will provide not only better flavors and nutritional qualities but special products. "We may see one bunch of farmers in Mississippi specializing in blue, shrinkproof cotton for denims, and another in Arizona supplying flameproof cotton for aircraft seats," says Calgene's Roger Salquist.

For scientists this truly is the golden age of bioengineering. Says Bill Belknap, who is developing the rot-resistant potato, "You only have to think of something and you can do it."

Cotton grower Sykes Sturdivant has gotten a glimpse of this brave new world. Farmers like him in the Mississippi Delta spray chemicals as often as 12 times a season, fighting bugs and weeds that kill more than a quarter of their crops. But at a research station in Alabama, Sturdivant saw a different future.

Rows of cotton plants had been sprayed only once with an herbicide. The weeds were dead, but the cotton kept growing. Even better, other rows, deliberately infested with cotton's most dreaded enemy, the cotton bollworm, showed no damage.

The first lot of "transgenic" cotton had been programmed to shrug off the herbicide, the second to manufacture a natural pesticide in its own cells. Similar traits are being engineered into soybeans, potatoes and corn. When the seeds of these "smart" plants come to market, crop yields should improve dramatically, and there will be less need for the million pounds of costly and potentially harmful pesticides farmers use each year.

Other environmental benefits may arise once plants are programmed to produce polymers, enzymes, pharmaceuticals and other industrial raw materials. "Fifty years ago plants provided almost everything we needed, and then petrochemicals took over," explains Chris Somerville of Michigan State University. "Now we're going back to plants, but with a lot of new tricks."

Somerville has produced biodegradable plastic by inserting a polymer-producing gene into a bacterium. The next step: transfer the gene into potatoes or sugar beets, and eventually churn out plastic for only 20 cents a pound. He believes it could happen within a decade.

Potatoes are even growing serum proteins for human blood transfusions. Normally extracted at high cost from human blood, the raw material of human serum albumin has been experimentally grown in greenhouses by scientists in the Netherlands and can be produced for as little as $16 a pound. "Nothing is cheaper than production in plants, because they consume only sunlight and water," says Peter Sijmons, head of the project.

Meanwhile, in North Carolina, tobacco infected with a harmless virus has produced experimental crops of Compound Q, a drug currently being tested for effectiveness against the AIDS virus.

"In theory there's no therapeutic protein we couldn't make," says Larry Grill of Biosource Genetics Corporation in Vacaville, California, which developed the procedure. . . .

From Thin Air

Bioengineering may change forever the way we farm. Scientists at the Plant Gene Expression Center in Albany, California, are illuminating the role genes play in a plant's reactions to light. "The research suggests different fields of the same crop might even be made to ripen in sequence, or farmers could delay blossoming if it turned out to be a cold spring," says Gerald Still, director of the center.

Biotech should also create plants that tolerate heat, drought, even salt. "The day is not far off when we will design plants to grow anywhere," says Jerry Caulder, president of Mycogen Corp. of San Diego. "Instead of changing the desert, we'll change the plants."

Another hope is that important crops can derive fertilizer from thin air. Beans and peas, aided by bacteria, extract nitrogen from the atmosphere and store it in their roots, but most crops draw it from the ground. At present, American farmers spend $12 billion a year on fertilizer, and half of it washes away or evaporates. Scientists from the United States and Europe are trying to get rice and corn to emulate peas and beans.

Unwarranted Fears

But the green-gene revolution will not get much further than the lab unless transgenic plants are accepted by consumers.

Concern about tampering with the genetic makeup of food is a powerful force. Because of initial public fears that "monster" plants or runaway weeds would be created, the release of transgenic plants has been rigorously controlled by scientists and a combination of federal agencies: the Environmental Protection Agency, the U.S. Department of Agriculture, and the Food and Drug Administration.

Both the National Academy of Sciences and the FDA say transgenic plants present no special risks, however. "No new technology has ever had so much oversight in place before the first product even appears," says Al Young, the Department of Agriculture's head of biotechnology.

The pioneering companies hope the benefits of genetically transformed products will be so tantalizing that any lingering resistance will melt away. The key point to remember is this: the green-gene technology that created a tastier tomato will also benefit the environment and help feed the 10 billion mouths that will be here within half a lifetime.

"Once it was science fiction to think of such things," says Jerry Caulder of Mycogen. "Now they're just around the corner."

> "No one really knows what effect splicing in . . .
> other genes into a plant would have."

Genetic Engineering Harms Agriculture

Joel Keehn

In the following viewpoint, Joel Keehn writes that genetic engineering's potential to improve agriculture has been exaggerated. He argues that large seed and chemical corporations are developing herbicide-resistant crops not to improve farming, but to consolidate their control of the agricultural industry. Keehn insists that these crops, and the chemicals on which they depend, pose various health and environmental threats, such as the creation of herbicide-tolerant weeds. Keehn is a freelance writer based in New York.

As you read, consider the following questions:

1. How might biotechnology exacerbate the world food situation, according to Keehn?
2. How does Wendell Berry, as quoted by the author, define sustainable agriculture? How does his definition differ from Robert Giaquinta's, as cited by Keehn?
3. What criticisms does the author make concerning the press coverage of genetic engineering issues?

Joel Keehn, "Mean Green," *Buzzworm*, January/February 1992. Reprinted by permission of the author.

When Rachel Carson launched the modern environmental movement a generation ago by taking the pesticide industry to task in her book *Silent Spring*, she thought she saw somewhere in the not-too-distant future a glimmer of hope—a time when farmers could throw away their chemicals and rely instead on plants biologically adapted to resist disease, weeds and insects.

The proponents of today's genetic engineering movement say that they have brought Carson's vision to life, creating in their laboratories a cornucopia of fruits, vegetables and grains whose genetic traits have been altered, heralding the birth of a brave new world for agriculture. Consider this advertisement from Monsanto, an early plug for their biotechnology program. The ad focuses on a single stalk of corn, growing in a desert. Underneath appears this message: "Will it take a miracle to feed the world?" The implication? Genetically engineered crops will appear like manna from heaven, crops will bloom in the desert, and world hunger will come to an end—a miracle of modern science bordering on the supernatural.

Such a vision, however, may be more illusion than reality. Despite the corn-in-the-desert image, neither Monsanto nor any other company is about to unveil crops able to survive drought or fix nitrogen and thus provide their own fertilizer source. Instead, one of the first products to emerge out of Monsanto's labs and into farmers' hands will be varieties of soybeans or canola able to withstand higher doses of one of the company's best selling and most powerful herbicides, Roundup—a move that, by at least one estimate, could boost worldwide sales of that product (a chemical that Monsanto has managed to extend its patent on) by $150 million a year.

Saving the world? Rachel Carson would be disappointed, but probably not surprised.

Supporting the Status Quo

Of course it's not just Monsanto that's working on herbicide-tolerant crops. And it's not just herbicide-tolerant crops that have environmentalists and people involved in sustainable agriculture concerned about the direction biotechnology is headed. "A lot of the early news about biotech crops has just been hype," says Dennis Keeney, director of the Leopold Center for Sustainable Agriculture at Iowa State University. "Most of the progress so far will support the status quo, high-input, industrial agriculture. Herbicide-tolerance is just the most obvious example of that."

The world's largest chemical and seed companies are all eagerly taking plants apart and genetically splicing them back together again, hoping to concoct a plant that can be doused with a herbicide powerful enough to kill all the weeds growing nearby but that won't bother the crop itself. A report from the Biotechnology Working Group, a group of scientists and citizens that monitors developments in biotechnology, shows that at least 27 corporations have launched research programs to develop herbicide-tolerant crops, tinkering with everything from soybeans and corn to poplars and petunias. The chemicals they are working on include some, such as atrazine and 2,4-D, that have contaminated groundwater supplies and are suspected carcinogens. Even the USDA [U.S. Department of Agriculture] has gotten involved, spending nearly $10 million since 1985 researching herbicide-tolerant crops.

Reprinted by permission of Ed Gamble.

Although people involved in sustainable agriculture are most immediately concerned about herbicide-tolerant crops, they worry also about some of the other developments coming from the biotech revolution. They worry that biotechnology could exacerbate the world food situation by continuing to place control of the world's genetic resources in the hands of a few transnational corporations. Others say that biotechnology will just maintain the American farmers' addiction to high-tech, capital-intensive farming that depends on monoculture crops like soybeans and corn—a practice that not only depletes soils and contaminates the land, but squeezes smaller, more diversified farmers out of business.

"I don't want to sound like a nay-saying, anti-technology zealot," says Charles Hassebrook, director of the Center for Rural Affairs in Walthill, Nebraska, who has studied the impact of genetic engineering on American agriculture. "But the areas consuming the greatest energy and money right now support industrial farming—growing the same crop in the same field year after year; applying expensive inputs to override natural systems; and benefiting large corporations while further eroding the role of the family farmer."

Defining Sustainable Agriculture

Of course, the companies working on genetically engineered crops have a different row to hoe—in part because they have different ideas about sustainable agriculture. "Sustainable agriculture does not necessarily mean reduced inputs," says Dr. Robert Giaquinta, manager of Du Pont's Biotechnology–Global Business section. "It means providing as much food to as many people as possible in as efficient a way as possible, while protecting the environment." Giaquinta believes that genetic engineering, like chemical fertilizers and herbicides before it, will essentially provide another high-tech tool to farmers allowing them to be more efficient.

Perhaps most notable about that definition is not only the relative unimportance it puts on the environment, but how it completely leaves out the farmers. Compare that to this simple definition offered by Wendell Berry, the Kentucky farmer, poet, and sometime social critic: "Sustainable agriculture is agriculture that does not deplete soils or people."

People like Berry, who include farmers and rural communities in their vision of sustainable agriculture, are often criticized for protecting a small group of folk who practice a quaint but anachronistic profession at the expense of an exploding world population that needs an ever greater and cheaper food supply. But Hassebrook disputes that perception, arguing that the move toward mechanistic, high-tech farming has actually jeopardized the world's food supply. Moreover, he argues that the main obstacle to increased farm productivity is rising energy costs, a problem exacerbated by a system that replaces human labor with expensive inputs. "The farming of the future is more vulnerable to shortages of energy than it is labor," he says. Hassebrook acknowledges that new technologies will play a role in agriculture's future, but believes that they should be used to understand and work with natural systems, rather than against them.

Companies involved in biotechnology argue that their current effort to alter the genetic makeup of crops isn't significantly different from what farmers have been doing for thousands of years. "People have been manipulating the genes of crops since the first farmer out in his field saw a plant he liked and saved its seed to plant the next year," says Roger Salquist, president of Calgene, Inc., the Davis, California–based company that is most aggressively trying to get its engineered crops to market. Indeed, nearly all of the crops grown in this country are the products of long classical breeding programs that have tried to emphasize beneficial characteristics in plants and weed out bad ones by crossing two or more varieties of the same species. For example, the fat red tomatoes we now eat have been developed over hundreds of years from a hearty but tiny-fruited and nearly tasteless tomato that grows in the South American Andes.

But classical breeding has several limitations. First, it's imprecise. For example, imagine you have a juicy tomato that is prone to disease that you want to cross with a tasteless, hearty one. You cross the two hoping to get a tasty, hearty tomato, but might end up with a tasteless, sickly one instead. Second, classical breeders can work only with plants within the same or closely related species. These limitations have prompted some seed companies to expose seeds to high temperatures, magnetic charges and even radiation in the hope that the seeds might randomly mutate, perhaps yielding new and interesting varieties.

A Powerful Tool

In comparison, genetic engineering provides a more versatile and powerful tool to biologists. "What we are doing now with genetic engineering is a faster and infinitely more precise version of what breeders have always done," says Salquist. "Instead of randomly growing plants to see which ones we like, we can decide what trait we want, find the gene we need from whatever plant or organism it is in, and then insert it directly into the crop we are working on."

Chemical and seed companies have long tried—and failed—to develop herbicide-tolerant crops through classical breeding techniques. Now that they've succeeded through genetic engineering, they say that herbicide-tolerant crops will allow farmers to move away from dangerous chemicals like 2,4-D and use newer, less toxic herbicides. For example, Calgene has developed a variety of cotton that can withstand applications of the chemical bromoxynil, made by Calgene's French partner, Rhone-Poulenc. Bromoxynil is less toxic to humans than is 2,4-D, and breaks down in the environment faster than atrazine, the most commonly used herbicide in the country. Because bromoxynil breaks down quickly, it is unlikely to leave residues in food or in groundwater. And because the chemical packs more plant-killing power per pound, farmers can reduce the amount of herbicides they use on cotton. Says Salquist, "Bromoxynil-safened cotton offers farmers an outstanding opportunity to save a bunch of money, and reduce their use of dangerous herbicides."

Monsanto makes a similar argument for its work on Roundup, or glyphosate, saying that it is an "environmentally benign" herbicide that quickly degrades. The company has conducted over 45 field trials for crops tolerant to glyphosate, including cotton, petunias, soybeans, sugar beets, tobacco, tomato and canola. "Life in the real world means that farmers are going to have to use some herbicides," says James Altemus at Monsanto. "This gives them a chance to be more flexible in choosing the ones that are effective and safe."

Birth Defects, Cancer Tied to Herbicides

But reports of the death of these older, more dangerous herbicides have been greatly exaggerated, says Rebecca Goldberg at the Environmental Defense Fund. "At the same time companies are working on glyphosate and bromoxynil, they're also working on tolerance to 2,4-D and atrazine," she says. In fact, atrazine-resistant canola and soybeans have already made it to the market in Canada. Atrazine has been found in the groundwater supplies of 12 states and 2,4-D has been linked to non-Hodgkin's lymphoma in farmers. The companies are working on these older, more dangerous chemicals because if farmers rely on only one herbicide, as they would be tempted to with herbicide-tolerant crops, weeds are likely to adapt quickly to them and the herbicide would no longer be effective—or profitable.

And while newer herbicides like bromoxynil and glyphosates might be less harmful than their older cousins, they still present some health risks. Laboratory tests show that some formulations of bromoxynil cause birth defects in animals and may pose developmental risks to the future offspring of people who apply the chemical. As a result, the U.S., Canadian, and California governments have restricted the chemical's use. And some of the so-called inert ingredients in certain versions of glyphosate are toxic to some fish and exotic organisms.

The risks of herbicide-tolerant crops in developing countries are particularly severe. Nearly all major crops grown in this country are not native here and so do not have wild relatives. But in South America and Asia, where crops such as rice, wheat, soybeans and potatoes all originate, wild relatives are common and frequently grow like weeds in farmers' fields. Scientists worry that in those countries herbicide-tolerant crops could cross-pollinate with their wild, weedy relatives and give rise to herbicide-tolerant weeds—adding a new and particularly troublesome nightmare to farmers there.

But perhaps most worrisome to people involved in sustainable agriculture is the way these crops will perpetuate American farmers' reliance on chemical inputs and continue to put control of agriculture in the hands of large corporations. "The big chemical companies know that this is the way for them to maintain their market share," says Margaret Mellon, who follows biotechnology for the National Wildlife Federation. "They'll sell the seeds and the chemical together as package deals."

But herbicide-tolerant crops aren't the only application of biotechnology that perpetuates the status quo in American agriculture—just the first.

Genetic engineering could possibly make plants resistant to common crop diseases, a development some environmentalists would welcome. But Hassebrook worries that seed companies, which sell primarily corn and soybeans, will focus on diseases that occur only when those two crops are planted year after year as monocultures.

Scientists are also trying to use biotechnology to fight insect pests. One approach capitalizes on the fact that some insects, such as certain wasps, help farmers by attacking pests. Following the herbicide-tolerance model, researchers at the University of California at Davis are trying to make these beneficial insects able to withstand commonly used insecticides. Farmers could then release these insecticide-resistant pest predators to patrol their fields at the same time they used chemicals also designed to kill the insect pests. The results would be the same as with herbicide-tolerant crops: continued reliance on pesticides, and a financial boost to the chemical companies.

Infatuation with Biotechnology

Another way to control insects uses a naturally occurring bacteria called Bacillus thuringiensis (Bt.), which contains toxins fatal to some insect pests but not to birds or humans. Scientists have transferred the Bt. toxin into varieties of tobacco, tomato and cotton. While the approach raises fewer questions than some other applications of genetic engineering, Rebecca Goldberg, at the Environmental Defense Fund, worries what impact the technique might have on food's nutritional quality. While the Bt. toxin does not affect humans, Goldberg says, "As high-tech as it is, genetic engineering is not an exact science. No one really knows what effect splicing in these toxins, or other genes, into a plant would have." Goldberg worries that these foreign genes could interfere with a food's natural nutritional value, or release new, potentially dangerous toxins.

Such criticism of biotechnology has largely been absent from the popular press, which for the most part has reflected the same infatuation with the new miracle technology as the Monsanto advertisement. Nor is that love affair limited to the press and industry. [Former] vice-president Dan Quayle issued a glossy report that praises biotechnology as the savior not only for American agriculture but the whole economy.

"The enthusiasm for this particular version of technological Utopianism has largely overwhelmed debate on the subject," says Jack Kloppenburg, Jr., professor of rural sociology at the University of Wisconsin. "But we should remember what people said about nuclear energy when it first came out, too."

But people who raise questions about biotechnology—who worry about herbicide-tolerant crops, the release of genetically engineered organisms, or its effect on foods—are often labelled as Luddites [people opposed to technological change]. Such an attitude has made it difficult for anyone—government policy makers, research scientists, farmers or consumers—to stop and consider where biotechnology is taking us and if that is where we really want to go. Or, as Dennis Keeney of Iowa State's Leopold Center puts it: "Everyone's so excited by their widgets and their widget-making technology, no one has stopped to see what good the widgets will do."

"The possibilities of agricultural biotechnology are enormous."

Biotechnology Improves Crops

Shari Caudron

Genetic engineering will enable farmers to produce crops that need little or no herbicides and pesticides and that are better for and more attractive to consumers, Shari Caudron argues in the following viewpoint. Caudron cites polls showing that most consumers favor biotechnology in agriculture and believe it will benefit them and lead to improved food quality and nutrition. Caudron is a writer and contributor to *Industry Week*.

As you read, consider the following questions:

1. What are some of the examples the author gives of crops improved through genetic engineering?
2. What are the three major concerns of the Pure Food Campaign, as cited by Caudron?
3. How is Calgene responding to the public's concern about genetically engineered crops, according to the author?

From "Supercows and Flounder Berries" by Shari Caudron. Reprinted with permission from *Industry Week*, December 6, 1993. Copyright, Penton Publishing, Inc., Cleveland, Ohio.

Some day, decaffeinated coffee may be grown on the vine. Citrus crops may withstand freezing temperatures. Cantaloupes may be immune to disease.

And if the promise of biotechnology holds, chickens will be bigger, pigs will be leaner, and cows will produce more milk. By tinkering with the genetic makeup of certain plant and animal species, scientists are creating bigger, better, stronger, and—in some cases—tastier sources of food. Depending on the "expert," the advancements made possible by genetic engineering will either solve world hunger or lead to dangerous biological pollution. Regardless of who is right, the brave new world of agricultural biotechnology is here to stay.

By 1993, the U.S. Department of Agriculture (USDA) had issued more than 461 permits to field-test genetically modified plants and microorganisms to such corporate giants as Monsanto, Sandoz, Du Pont, and Upjohn, as well as to biotech upstarts such as Calgene and DNA Plant Technology. Among the plants currently undergoing some type of genetic testing are cotton, tobacco, potatoes, corn, soybeans, tomatoes, and melons.

Manipulating Genes

With genetic engineering, "bad" genes can be turned off, "good" genes can be enhanced, and desirable characteristics from plants, animals, and even humans can be transferred from one species to another to create new "transgenic" foods. Looking to grow potatoes that resist fungus? Tap into the genetic structure of the pea, which is known for its fungal resistance, isolate the gene that allows it to resist infection, and transfer it to the potato. Voila! You now have a new fungus-resistant potato. Genetic engineering actually has been going on for as long as people have been breeding plants and animals. The mule is an artificial genetic cross between a donkey and a horse. The difference between the genetic engineering of yesterday and today is that breeders used to use whole animals and plants to create new crossbreeds. For example, conventional crossbreeding produced an early maturing peach, but thousands of other genes transmitted by the breeding caused the peach to crack open on the tree. Now, scientists have the ability to identify and select only the genes they want, allowing them to create a mature peach without all the unwanted side effects.

Considering the staggering number of individual genes, the possibilities of agricultural biotechnology are enormous. Scientists at Monsanto in St. Louis, Missouri, for example, have created a genetically altered high-starch potato that absorbs less oil during processing, meaning less fat and calories for french-fry and potato-chip lovers. A hormone that boosts milk production in dairy cows [was approved by the] Food and Drug Administration

83

(FDA) in November 1993. Wheat and other grains are being manipulated to create their own nitrogen, thus reducing the need for fertilizer. And ESCAgenetics, in San Carlos, California, is working on a variety of sweet corn that has a two-week shelf life, instead of the two-day life common today.

The first genetically engineered product to hit the market is the FLAVR SAVR tomato developed by Calgene, of Davis, California. The company was able to locate the gene that causes tomatoes to soften and spoil rapidly. By removing the gene from the tomato's DNA, reversing it, and reinserting it, the company has created a tomato variety with a softening process slowed by about 90 percent. This means that tomatoes will be allowed to vine ripen before being shipped to stores. If everything is timed just right, tomatoes will arrive on the shelf fat, red, and sweet, just like grandpa used to grow 'em. . . .

Calgene isn't the only company to target the pathetic state of store-bought tomatoes. In fact, the USDA has given more permits for testing on tomatoes than any other crop. DNA Plant Technology Corp. (DNAP), of Cinnaminson, New Jersey, for example, has created a tomato designed to deliver "shelf-life extension." The tomato will stay ripe on the vine for up to two months and will stay ripe for eating three to four months after picking. . . .

Educating Consumers

Biotechnology companies and consumer activists have waged a pitched battle over the future of the genetically engineered cow-growth hormone, bovine somatotropin (BST). On November 5, 1993, the Food & Drug Administration finally approved Monsanto Co.'s controversial drug, . . . which boosts milk production in cows. . . .

Monsanto is trying hard to counter arguments against the drug. It is sending informational packets to thousands of dairy farmers, as well as meeting with them. It also has lined up 250 doctors and other health experts nationwide to answer consumer questions about health issues, and it plans to provide educational brochures to retailers.

David Greising, *Business Week*, November 22, 1993.

The tomato is receiving so much attention because there is no other produce item that is as likely to leave consumers dissatisfied as the tomato. Yet, the fruit is eaten in 85 percent of all households, and the U.S. tomato market is worth between $3.5 billion and $4.2 billion a year. The company that successfully markets truly tasty tomatoes will score big in the profit department. DNAP, which has spent between $2 million and $4 million developing its VineSweet brand, could recoup its costs in anywhere from six months to three years.

Admittedly, the possibilities of genetic tinkering can get pretty strange. For example, human genes can be added to pigs in order to create leaner meat, and a gene that allows arctic flounder to survive in frigid waters might also allow oranges and strawberries to survive frost damage in the field. Plants can incorporate fish genes and vice versa, because all living organisms read the same genetic language. But don't look for pork chops hanging from fruit trees anytime soon. Genetic research is expensive and time-consuming, and success for the companies involved will depend almost entirely on consumer acceptance of the products they create.

Currently, the United States leads the world in genetic-engineering research. There are 602 U.S. companies listed in the 1992 *Genetic Engineering News Guide to Biotechnology Companies*, while the 212 remaining world biotechnology firms are divided among 25 countries. Biotechnology was a $6 billion business in 1992 and is expected to reach $50 billion by the year 2000, says Dick Godown, senior vice president of the Biotechnology Industry Association in Washington, D.C. Of this, only about 10 percent is devoted to agriculture. "But because we're dealing with commodities," Mr. Godown says, "we'll be talking significant numbers when these products start to hit the market."

Considering the Risks

Because agricultural biotechnology holds such great economic promise, in 1992 the White House Council on Competitiveness succeeded in encouraging the FDA and the USDA to restructure their positions on biotechnology products. Instead of subjecting genetically engineered foods to a set of rigorous tests, the new policy does not treat these new foods any differently from natural or traditionally bred foods.

As Marvin Scher writes in an article in *Food Processing*,

> The new FDA and USDA policy abandons the concept that processes like gene splicing are inherently dangerous. Instead, the focus is on the product itself and the environment into which the product will be introduced. Thus, a gene taken from a tomato, reversed and reinserted, seems unlikely to cause harm, and the product stands ready for market without restrictive documentation. On the other hand, a species of fish newly equipped with large dentures could disrupt the ecosystem if released. This would require greater scrutiny and regulation. . . .
>
> Regulators will intervene only when the risk is unreasonable. . . . They appear to be more willing to judge risk relative to benefit rather than to arbitrarily set risk at a level near zero.

But Dan Barry, director of the Pure Food Campaign (PFC), an

organization opposed to genetically engineered foods, isn't buying it. "Genetic engineering rewrites the blueprint of life, creating new species only because they have traits beneficial to humans. Do we as a species have a right to do this?"

The PFC, which is supported by 2,000 chefs nationwide, has three major concerns with these new "frankenfoods." One, that the genes added to plants or animals could affect other genes and create foods that are highly allergenic. A person who is allergic to peanuts, for example, may be affected by foods that have been altered with peanut gene. Two, the organization believes that the release of genetically engineered organisms into the environment could lead to the biological pollution of native plant and animal species. And three, the campaign thinks that the failure of the FDA to label genetically engineered foods means that persons who follow religious dietary restrictions will be unable to ensure compliance with their beliefs. A pork gene inserted into zucchini, for example, may be unacceptable to some Jews.

In response to the advent of genetically engineered products, the PFC has launched an international campaign to boycott genetically engineered foods. The campaign is urging Americans to contact the FDA to demand mandatory premarket safety testing of all genetically engineered foods, mandatory labeling of the entire contents, and registration of new foods with the FDA so that they can be traced if illness or other problems arise.

Improving the Quality of Life

This kind of controversy presents challenges to the companies that are hoping to bring new foods to market. One of them, it appears, lies in educating consumers about the safety of genetically engineered foods. Because its FLAVR SAVR tomato is the first genetically modified whole food to hit the market, Calgene thought it prudent to subject the tomato to the FDA's full regulatory-review process, even though it was not required. "We wanted to enable the consumers to participate in the regulatory process and to give them an opportunity to see that safety is not an issue," explains Stephen Benoit, vice president of Calgene Fresh, of Evanston, Illinois.

Furthermore, the company has voluntarily decided to provide at the point of purchase educational brochures that explain the technological process used to develop the tomato. "What this comes down to is [whether we] can deliver a better-tasting tomato to consumers," Mr. Benoit says.

Jim Altemus, manager of plant-science communication at Monsanto Agricultural Group in St. Louis, adds that the capacity of biotechnology goes well beyond satisfying the consumer's need for tastier produce. Biotechnology can help reduce our reliance on chemical pesticides, he says, slowing the depletion of the ozone layer. Because genetically engineered foods have greater resistance to insects, disease, and drought, they also can help solve the crop tragedies experienced by farmers in Third World countries. "I don't think we have the right," Mr. Altemus says, "to deprive others of technology that can improve their quality of life."

Consumers are cautiously optimistic about agricultural biotechnology. According to a survey conducted by Professor Thomas Hoban at North Carolina State University, two-thirds of consumers believe they will personally benefit from biotechnology, and almost 75 percent believe that biotechnology will have a positive effect on food quality and nutrition. However, this doesn't mean consumers will accept every new creation thought up by bioengineers. There is lower acceptance of biotechnology when it is used to change animals compared with plants, and survey respondents seem unwilling to accept gene transfers between plants and animals or between humans and farm animals. In other words, moving a gene between tobacco and broccoli is one thing, but inserting human genes into chickens is different.

"We need to know why biotechnologists suppose they can avoid producing the negative impacts comparable to those created by chemical and nuclear technologies."

Biotech Agriculture Is Inefficient and Destructive

Wes Jackson

Wes Jackson is the director of the Land Institute, an educational and research organization devoted to sustainable agriculture, located in Salina, Kansas. Author of the book *New Roots for Agriculture*, he is well known as an environmentalist and naturalist. In the following viewpoint, Jackson maintains that the "progress" of biotechnology is simply a continuation of humankind's blindness to how nature works and what is good for the environment. He believes that naturally occurring processes are more efficient and environmentally sound than new agricultural technologies such as genetic engineering.

As you read, consider the following questions:

1. Why is the worldview of biotechnologists worrisome, according to Jackson?
2. What is Jackson's "evolutionary-ecological" worldview?
3. According to the author, why is the concept of "smart" resource management flawed?

Wes Jackson, "Listen to the Land," *The Amicus Journal*, Spring 1993. Copyright ©1993 by *The Amicus Journal*, a publication of the Natural Resources Defense Council, 40 W. 20th St., New York, NY 10011. Reprinted with permission. NRDC membership dues or nonmember subscription: $10 annually.

How long did it take chemists to come up with a substance that would destroy the ozone layer? Whether the answer is a century or half a century matters little. The point is, the chemists finally did it. The question now is, "How long will it take the biotechnologists to come up with something that is as disastrous as the ozone hole?" Surely they will not let the chemists hold the record. I am being sarcastic because it seems nearly impossible to get the scientific establishment to think about the very basic assumptions under which most of us scientists operate.

Modern science operates under the assumptions given us by seventeenth-century philosopher and mathematician Rene Descartes, who admitted, in his *Discourse on Method*, that the more he sought to inform himself, the more he realized the magnitude of his ignorance. Rather than regard this as an apt description of the human condition and the very proper result of a good education, however, Descartes thought his ignorance correctable. The knowledge-as-adequate world view was born.

I have heard grown men and women argue whether it was safe or prudent to release ice minus bacteria, the genetically engineered microbe designed to keep frost off crops. Meanwhile, the Beltsville hog—arthritic, cross-eyed, suffering from renal disease and an enlarged heart, the kind of hog any self-respecting farmer would knock in the head—carries a human gene that is supposed to make it gain faster and be leaner. Some gene splicers will say that what the hog needs is some fine-tuning—they clamor for more research. I am less concerned about this hog monster than the human monster, created by our culture, the monster who sees nothing wrong with creating such a hog.

A Disastrous World View

Even more worrisome than the question of the environmental and health risks implicit in biotechnology are the inherent moral questions surrounding the Cartesian [after Descartes] world view, and, by extension, the world view that the biotechnologists are advancing. This world view has already bred a host of disasters, from Chernobyl [the 1986 explosion at the Chernobyl nuclear power plant in Ukraine] to the spill of the *Valdez* [the 1989 spillage of nearly 11 million gallons of crude oil into Prince William Sound, Alaska]. One has only to introduce one variable, the proverbial drunken sailor, and one can predict a stochastically [randomly] assured accident.

We scientists are the ones who have created these different times, but humanity, back to the Greeks at least, has understood the problem of *hubris*. The Greeks understood that hubris, or overweening pride, means introducing a human-made pattern into the world, one that disrupts a larger pattern. What we must understand is that the larger pattern is not of our making and we must accept that we depend upon it.

In *The Death of Ramon Gonzalez: The Modern Agricultural Dilemma*, Angus Wright reminds us that chemical companies manufacture about 5 billion pounds of lethal substances every year, designed to be cast into air, soil, and water. In spite of our concern with the disposal of toxic waste, "corporations, governments, and international agencies encourage the deliberate dispersal of this fantastic quantity of biocides into the environment," he writes.

By permission of Chip Bok and Creators Syndicate.

Wright explains how this became routine: "For very large conspiracies to work over large geographical areas and for decades at a time, the conspiracy must be transformed into something else—a belief system, an ideology, a world view, a concept of proper professional behavior, even a crusade." The world view that makes agri-chemicals inevitable makes the following assumptions: 1) nature is to be subdued or ignored, 2) the purpose of agricultural research and farming is to increase production, and 3) agriculture is to serve as an instrument for the advancement of industry. When we made CFCs [chlorofluorocarbons, which some blame for a depletion in the earth's ozone layer], we *ignored* nature.

The Industrial Revolution led inevitably to the Third World's exploitation of labor and chemical assault on its peoples, its land, soil, and water; these are consequences of what Wright calls "the modern agricultural dilemma," where nations rapidly industrialize to expand their agricultural exports and join the international economy. This expansion, however, jeopardizes the "highly localized adaptations needed for ecologically healthy agriculture and healthy, stable rural communities"; it also has far-reaching environmental and health effects. "Booming agricultural regions like the Culiacan Valley [in Mexico] actually depend on the cheap labor produced by ongoing social and environmental disaster in other regions," Wright points out, "and worse, the technologies being used to produce present growth make it likely that the new regions . . . will follow the old into social and environmental decline."

A New World View

We should now adopt an operating paradigm, I believe, that is ecological in nature; because it is ecological it is also evolutionary. Time-honored arrangements are on our side. We might call the new world view an evolutionary-ecological world view and use it to inform our decisions.

A more effective goal for all of us is to get serious about becoming "native to our place." As a culture, we still operate more in the conquering spirit of Columbus and Coronado than in that of the natives we conquered. To be native to this place would not mean the end of science or the end of management of our landscape.

To illustrate this goal, I have two stories: In 1933, a graduate student at the University of Nebraska compared an upland, never-plowed prairie with a field of winter wheat. Prairie and wheat field grew side by side in the same soil. When the rain fell, 8.76 percent ran off the wheat field, but only 1.2 percent ran off the prairie. The wheat plants died, but the deep, penetrating roots of the prairie allowed the grass to survive. The upshot: native prairie was designed by nature to receive water efficiently and to allocate it carefully.

Far from Nebraska, in a tropical rain forest in Costa Rica, Jack Ewel and his colleagues from the University of Florida have compared agricultural fields with the surrounding forest. Here, water is a nemesis of fertility, for when the forest is destroyed, valuable nutrients are leached downward. A forest, in contrast to a prairie, is "designed" to pump water back into the atmosphere with great efficiency.

Here are two very different ecosystems with respect to water management, one "designed" to hold water, one to get rid of it— but each is keyed to its place. In each case, too, whenever humans interfere, we tend to invert what nature does well. The poet Alexander Pope understood that, and advised "consult the genius of the place in all."

Joining Ecology and Economics

At the Land Institute in Salina, Kansas, we have consulted the genius of the place and have devoted lots of time, thought, and resources to using nature—in our case native prairie—for biological agriculture. At various sites on our 100-acre, never-plowed prairie, ecologist Jon Piper and students sort above-ground plant material into groups of cool and warm season grasses, legumes, sunflowers, family members, and other. Knowing how the ratios of these plant groups differ across soil types and in wet and dry years, and using data derived from soil-root interaction, we have set out to roughly mimic native vegetative structure.

Ewel and his colleagues have also mimicked native succession in the tropics. By substituting vine for vine, tree for tree, shrub for shrub, using plants requiring human intervention only, Ewel concludes that nearly always when the native vegetative structure is imitated, the result is high productivity, responsiveness to pests, and good protection of the soil.

The fusion of ecology and agriculture has produced a hybrid discipline called agroecology. The human break with nature came with agriculture, so it is fitting that agroecology arrange a long-term courtship between ecology and economics. But before there can be a marriage, the current world view of the run-of-the-mill economist, like that of the run-of-the-mill agriculturist will have to change.

Some small progress has been made. A few agricultural researchers are now willing to consider the evolutionary history of domesticated livestock. Once they acknowledge that domesticated plants and animals are relatives of wild things that evolved in a context not of our making, they will make a fundamental shift in their thinking. They will start to see the chicken, for example, as fundamentally a jungle fowl, not as a piece of property that produces eggs or meat when confined in a small cage. The ecological perspective honors jungle fowl, forest animals, and savanna grazers (the ancestors of beef and dairy cows)— and in the future, I hope domestic prairie seeds. As we move toward the larger notion of relationship, the word "resources" becomes obscene.

Yet nearly all of the research into and practice of sustainable agriculture is devoted to "smart resource management"—the notion that humans are getting smarter. Instead, we should be drawing the intelligence embedded in natural ecosystems up to us; this would go far beyond whatever "smart" capacities humans can muster. We need to rethink the assumptions of science and technology, and integrate ethics, poetry, and spiritual dimensions into our fields; we need to explore the connections among them.

For a while, at least, scientists should stop trying to solve specific contemporary problems and adopt generic ways of thinking that science brings to every field. Properly done, science will become more lively, more truly imaginative, more sensitive to real needs. We should imagine ourselves as explorers whose goal is to become native to this place. We would transcend ordinary ideas about sustainability, for we would be trying to live within our means in all those little places and with a cultural richness that transcends economics. This does not mean that we would leave science and technology behind.

Disease resistance and nitrogen fixation through gene splicing will probably work, just as chlorofluorocarbons worked in refrigeration. But the unforeseen and unforeseeable consequences of this world view is what is on the line. Had we been less willing to accept the mantra of the petrochemical age—better living through chemistry—it is likely that we would not be in our current fix with toxic waste and a growing ozone hole.

The Failings of Human Cleverness

Biotechnology is the current extension of the human cleverness approach to arranging the world. My objection is not that it is new but that it is painfully old—it is "writing the past large." These multimillion-dollar monuments to the gods of human cleverness are false gods, ones that we will one day be embarrassed to have worshiped.

Before biotechnology goes any further, we need to know why biotechnologists suppose they can avoid producing the negative impacts comparable to those created by chemical and nuclear technologies. How do we act on the fact that we are more ignorant than knowledgeable? Embrace the arrangements that have shaken down in the long evolutionary process and try to mimic them, ever mindful that human cleverness must remain subordinate to nature's wisdom.

"The production of transgenic food animals will . . . improve the quality of life for both animals and humans."

Genetically Altered Animals Will Benefit Humankind

Duane C. Kraemer

Duane C. Kraemer is associate dean for research and graduate programs at the College of Veterinary Medicine at Texas A&M University in College Station. In the following viewpoint, Kraemer contends that genetic engineering could make farm animals healthier and better adapted to farm environments. In addition, according to the author, such animals could potentially increase the availability and improve the quality and safety of food for human consumption. Kraemer concludes that the ethical issues raised by genetic engineering should elicit caution, but should not be allowed to derail scientific progress.

As you read, consider the following questions:

1. What two major advances must occur for transgenic food animals to be developed, according to Kraemer?
2. How might transgenic food animals improve food quality, in Kraemer's opinion?
3. Some critics oppose altering animals genetically because of the unforeseen consequences. How does the author respond to this argument?

Excerpted from "The Benefits and Ethics of Creating Transgenic Food Animals." This article appeared in the December 1994 issue and is reprinted with permission of *The World & I*, a publication of The Washington Times Corporation, ©1994.

One of the most powerful tools for the improvement of human and animal well-being is the use of recombinant DNA technology. This tool can be used in many ways; however, this viewpoint will be limited to that portion of recombinant DNA technology that involves the production of transgenic food animals.

Transgenic food animals are those to which genetic material containing DNA has been transferred intentionally to modify the genome (hereditary factors). This process is usually referred to as gene transfer, genetic engineering, or transgenesis. Gene transfer, as applied to food animals, is generally germ-line gene transfer, which means that the transferred gene can be passed on to future generations through the germ cells (spermatozoa or ova). As applied to humans, gene transfer is for correction of genetic defects and involves the transfer of genetic material to somatic cells (cells other than germ cells) such as those in the bone marrow.

For purposes of this discussion, food animals are those that produce meat, eggs, milk, or honey that is consumed by humans for nutritional purposes. Although the most common food animals are livestock (cattle, swine, sheep, and goats), poultry (chickens, turkeys, geese, and ducks), and fish, it should be realized that humankind consumes, intentionally or unintentionally, almost every form of animal life, both domesticated and wild. Some of the less common food animals include horses, deer, ostriches, dogs, cats, rodents, squirrels, snakes, turtles, lobsters, shrimp, crab, worms, and crickets.

Consumption of Transgenic Foods

Would you eat food produced by a transgenic animal? That is a question that you have not yet had to answer. In fact, that issue will most likely be resolved regarding food from plants several years before food from transgenic animals becomes available. Few people are aware that most of the cheese we consume is made with an animal enzyme, rennin, that is produced by transgenic bacteria.

To consume or not to consume food from transgenic animals is not the only issue. Some of us are considering the question of whether or not transgenic food animals should be produced, and, if so, what precautions should be taken to avoid undesirable consequences. This viewpoint discusses the potential benefits of producing transgenic animals and considers the ethics of such activities.

It is difficult to separate these two issues. What one person considers a benefit, based on a certain set of ethical assumptions, might be considered a detriment based upon a different ethical point of view. There are at least two groups of people who would likely reject the idea of producing transgenic food animals without further consideration: those who consider it unethical to produce transgenic animals for any reason and those who do not intentionally eat meat. Some members of the latter group might approve of the use of transgenic food animals for the production of medications or as models for research on human disease. Certainly, I consider both of these to be benefits as long as they can be achieved within the limits of appropriate animal care and use.

Potential Benefits

Here I use the word *potential* because, although transgenic cattle, pigs, sheep, goats, chickens, fish, and rabbits have been produced, none are being propagated for food production to my knowledge. Only those strains that are producing medications in their milk are being propagated for commercial purposes. Prediction of when transgenic food animals will be available is very difficult. It depends, of course, on the extent to which resources are dedicated to this effort. Two major advances are needed: (1) increased understanding of the genes that influence animal production and well-being, as well as the quality and quantity of food produced; and (2) improved efficiency of production of transgenic animals with predictable modifications.

The production of transgenic food animals will serve mainly to improve the *quality of life* for both animals and humans. As improvements in quality of life are achieved, quantity of life, expressed as greater longevity, is also likely to improve. More specifically, animal well-being might be improved by increasing their resistance to infectious diseases and increasing their adaptability to production environments. Human well-being might be improved by increasing the availability of food and improving the quality and safety of food animal products.

An example of this type of benefit would be the development of cattle that are resistant to zoonotic (those affecting both animals and humans) diseases, such as tuberculosis and brucellosis. This could probably be accomplished by the transfer of genetic material derived from naturally resistant cattle, some of which have already been identified. In other cases, it may be necessary to transfer genes derived from other species that are resistant to specific diseases. Increasing the resistance of food animals to diseases would clearly benefit the animals by reducing illness as well as the frequency of handling for such things as vaccinations, diagnoses, and treatments. Humans would benefit from reduced exposure to the zoonotic diseases, reduced exposure to antibiotics and other medical residues, and reduced costs of food production, which means lower prices for food and/or increased profits for producers. Having animals with increased resistance to disease would also permit the production of cattle in additional areas of the world where they cannot be produced effectively now. By making more effective use of natural resources and reducing inequities in food supply, famine and political unrest could be reduced or eliminated.

Genes Common to All

There is probably less ethical concern for the use of gene transfer to increase the disease resistance of animals than for any other of the proposed uses. This is especially true if the genes can be derived from the same species to which they are to be transferred. However, the source of the DNA is not necessarily a clear-cut issue. Since the chemical bases are identical across all species and can be exchanged between all species, does it make any difference where they are derived? The factors that make species unique are the products of the genes. These are the proteins that the cells produce by utilizing the gene as the template or the "pattern" (often called *gene expression*) and, indirectly, other molecules that the cells produce utilizing these proteins in the form of enzymes, cofactors, and other regulators of cell function. Most of the genes that make up the various species of mammals are common to all, with only a relatively few being unique to any single species. Some people have objected to the transgenic use of genes derived from human cells. Probably they do not realize that we ingest human DNA and cells when we swallow, eat food prepared by humans, or kiss one another.

Asexual Cows Beneath Cloned Trees

Cow embryos are made in the laboratory by fertilizing desirable eggs with superior sperm, allowing them to divide and splitting them into smaller portions which are introduced into new mothers (who need not themselves have any particular merit). This multiplies the number of high-quality calves. It is easy to freeze the embryos until they are needed and surrogate motherhood is already used on thousands of cows each year. . . . The rural landscape may become one in which asexual cows feed on engineered grass under the shade of clonal trees.

Steve Jones, *The Language of Genes*, 1993.

Increasing disease resistance is just one of many ways that genetic transfer could make animals more adaptable for food production. Animals could be made more adaptable to various temperatures, altitudes, or light-dark cycles, for example. It is clear that each of these characteristics is regulated by genes and that there is a natural variability within and among species for these traits. Therefore, gene transfer would very likely be able to create these changes.

Another approach to improving the adaptability of animals for food production is through behavior and/or sensory modification. Many different options could be considered, and some are controversial. Part of the controversy comes from the assumption that these modifications would be intended to adapt animals to conditions of close confinement and to lower their sensitivity to the stresses of such environments. Although this is certainly one option, the animals could be modified to make them more manageable under conditions of less confinement by being made easier to gather from rough terrain. Those persons who object to the killing of animals for food on the grounds that some animals are aware of their fate might prefer that the animals be modified to be without this capability. For some people, this would be the worst possible use of this technology. As stated above, some people's benefits are other people's fears.

Improving Food Quality

Another potential benefit from the production of transgenic food animals is the improvement of food quality. Knowledge of health-related issues concerning quantities or qualities of fat (lipid) in the human diet has prompted producers to search for animals that produce foods containing both less and higher quality fat. Such modifications in animal fat will be complex because it will be important to maintain such qualities as flavor and tenderness, which are often positively associated with lipid content and character. Efforts are under way to identify genes responsible for flavor and tenderness, in the hope that these characteristics can be transferred to animals that have desirable quantities and qualities of fat.

Food producers in free-enterprise economies are constantly striving for increased efficiency of production. Although this may appear to be a selfish effort to increase profits, it generally translates into greater availability of food at reduced cost to the consumer. There are numerous ways that transgenic food animals might increase the efficiency of food production. Transferred genes could increase the ability of animals to consume and convert recycled or waste products to useful and healthful foods even more efficiently than they do at the present time. Reproductive efficiency might be enhanced by transferring genes that would increase ovulation rate or decrease embryo and offspring mortality as well as birthing intervals. Changes in metabolism might be induced to achieve more efficient feed conversion, more rapid growth, or changes in mature body size. Any of these changes would likely result in greater production efficiency.

98

One of the major concerns expressed by those who oppose the use of transgenic animals for food production is their impact on the world's ecology. Indeed, assessing impact is a logical concern regarding any new or emerging technology. As with most other technologies, gene transfer in food animals can be either a benefit or a detriment to our environment. Some potential ecological benefits have already been mentioned, such as creating animals that can utilize waste materials and adapt to ecologically sustainable environments. Other ecological benefits might include reducing the number of animals needed to produce enough food; altering animal excretions to a more ecologically advantageous form; reducing the need for critical nutrients such as water or high-quality protein; changing animal behavior to conform to reduced space requirements; or altering grazing habits to reduce erosion.

A major ecological concern is the transmission of unwanted genes to the existing animal population. In species such as cattle, unwanted breeding is easily controlled. However, with species such as fish or snails, control is much more difficult and might require sterilization of the transgenic animals before they are released into the general environment. . . .

Gene Selection

Just because it is possible to integrate DNA from any source into a genome does not mean that every gene product will be beneficial for the animal or its products. Certainly there are many more genes that could be transferred than would be beneficial. Therein lies one of the most difficult problems in this enterprise—identification and use of the genes that will result in beneficial changes. Gene selection is generally based upon the effect of the gene product on the donor animal and on tissue culture cells. Beyond this, gene selection is a trial-and-error process. As a result, some transgenic animals produced are less healthy than those without the transferred gene.

Thus, one is faced with an ethical dilemma. Is the benefit that might be obtained from the gene transfer worth the risk of producing an unhealthy animal? The risk can be mitigated by:

• production of only a few transgenic animals at a time from any given gene construct;

• humane euthanasia of any defective offspring. (In many cases these defective offspring will be valuable for biomedical research, but this use would be unacceptable to people who disapprove of the use of animals for such research.) . . .

Additional Issues

I would like to discuss some concerns about consumption of products from transgenic animals. There is concern about consumption of recombinant DNA. In this regard, there is no evidence that consumption of recombinant DNA or any other cellular DNA has undesirable effects on the consumer. DNA is rapidly degraded during processing and consumption. We consume approximately two milligrams of nucleotide (DNA and RNA) daily in our normal diets from a wide variety of sources. This occurs without changes being induced in our genomes. . . .

Please understand that the benefits and ethical considerations pointed out so far are based on the current state of the art and developments that are anticipated in the near future. It is expected that there will be continued refinements of the technologies involved in the production of transgenic food animals. Also, procedures will be developed that cannot even be imagined at this time. What may be seen as ethical problems today, such as the low predictability of results of gene transfer, will very likely be solved by new knowledge. But, of course, each of these discoveries will bring with it a new set of ethical dilemmas.

The Ethics of Animal Care

Probably the most complex, and certainly the most controversial, ethical issue is that of producing food animals in a humane manner. It is fascinating that people from quite similar religious and cultural backgrounds may reach considerably different conclusions regarding the ethics of animal care and use. Some people conclude that humans are unique and have dominion over all of the animals. Within that group are those who interpret this to mean that there are virtually no limits to what may be done to animals. Others, however, interpret dominion to infer responsibility for animals, concluding that the killing of animals for food is ethically unacceptable. In addition, there is a spectrum of beliefs that fit between these two extremes. Some people believe that all animals are sacred, even entitling them to privileges not afforded to most people. Still other people consider some animals to be sacred but others to be totally exploitable. Thus, it is not surprising that there is no single set of rules that can be applied to develop a unified ethical conclusion regarding the creation of transgenic food animals.

Some of the most vexing ethical issues relate to the long-term effects of today's decisions. Some people maintain that we should not produce transgenic food animals because we cannot predict the long-term ecological or sociological effects of this practice. Certainly we must be concerned about long-term effects of our decisions. However, we can no better predict the long-term consequences of *not* producing transgenic food animals than we can predict the long-term consequences of producing them. I submit that our main long-range concern should be to provide for as much genetic diversity as possible, along with as many options for use of that diversity as can be maintained.

There might be some truth in the belief that humankind would have been better off without modern technologies. However, we do not have the luxury of starting over again as a species. Therefore, we must use our intellect and other resources to improve and sustain the world around us. It is my belief that transgenesis is a procedure whose time has come. It is our responsibility to use it wisely. Clearly, that will require patience, creativity, and compassion.

"Genetic engineering not only causes great animal suffering but also puts the genetic integrity of many species of animals at risk."

Genetically Altering Animals Is Dangerous and Inhumane

Andrew Kimbrell

Recent efforts at genetic engineering have produced horribly deformed pigs and other animals, Andrew Kimbrell writes in the following viewpoint. Kimbrell believes it is unethical to produce such creatures. In addition, he argues that genetic engineering could decrease animal diversity and threaten public safety. Kimbrell is executive director of the International Center for Technology Assessment in Washington, D.C., and the author of *The Human Body Shop*.

As you read, consider the following questions:

1. What are some of the genetic engineering projects being attempted in Great Britain, Australia, Canada, and the United States, according to the author?
2. How could genetic engineering lead to the further spread of AIDS, in the author's opinion?
3. How are animals genetically altered to produce human biological materials, according to Kimbrell?

Andrew Kimbrell, "Facing the Future: Genetic Engineering," *The Animals' Agenda*, January/February 1995. Reprinted with permission.

Pig No. 6707 was designed to be a super pig—super fast growing, super big, super meat quality—one of the first of a series of high-tech animals that would revolutionize agriculture. In 1988 researcher Vernon Pursel and his colleagues at the United States Department of Agriculture (USDA) research center in Beltsville, Maryland, had injected the human gene that governs growth into No. 6707 while he was still an embryo. The gene was supposed to become part of No. 6707's genetic code and thus help to create a pig that would grow larger and faster than any other pig had grown before.

Pursel had hoped that his super pig would mimic the success of the super mouse that had been created in 1982 by Ralph Brinster, a genetic engineer at the School of Veterinary Medicine of the University of Pennsylvania. No. 6707 did not turn into a super pig, however. The human growth gene altered his metabolism in an unpredictable and unfortunate way, and he turned into a tragicomic creation: an excessively hairy, lethargic, apparently impotent, slightly cross-eyed pig, riddled with arthritis, who rarely stood up. The USDA tried to rationalize its experiments by noting that even though genetically altered pigs were not larger than ordinary pigs, their meat would be leaner because of their greater muscle mass. It was obvious, nevertheless, that No. 6707 was the wretched product of a science without ethics.

Each year tens of thousands of experiments are conducted around the world by numerous researchers who add, delete, recombine, edit, and insert genetic material across biological boundaries, creating plants and animals that have never existed before. Many researchers, following Pursel's example, engineer human genes and those of other species into livestock and poultry in order to create super, more efficient animals for slaughter and other food production. Ten thousand pig embryos in Great Britain are injected with foreign growth genes annually. Australian researchers, having produced genetically engineered sheep that grow 30 percent faster than normal ones do, are currently transplanting a variety of genes into sheep in an attempt to make their wool grow faster. Researchers in Canada have experimented with boosting salmon size by inserting human, chicken, and cattle growth hormones into the fish; and researchers at the University of Wisconsin have managed to overcome mother hens' natural, "inefficient" brooding instinct so that the hens will spend more time laying eggs and less time sitting on them.

U.S. Efforts

Since the mid-1980s the American government and private researchers have expended billions of dollars, much of it provided by taxpayers, in an effort to produce "transgenic" animals that contain the traits of humans and other species.

Rodents and livestock containing human genes are commonplace at several U.S. corporate, university, and government laboratories. Carp, catfish, and trout also have been engineered with a number of genes from humans, cattle, and rats to increase the fishes' growth and reproduction. Researchers at the University of California at Davis used cell-fusion techniques to create "geeps," astonishing goat-sheep combinations that have the faces and the horns of goats and the bodies of sheep. In the early 1980s U.S. genetic engineers isolated bovine growth hormone (also known as bovine somatotropin), a naturally occurring hormone that stimulates cows' milk production. Scientists discovered that dairy cows injected with the genetically engineered hormone produced up to 20 percent more milk than nontreated cows did. Injections of the hormone also created immense animal suffering, including a massive increase in udder infections, digestive disorders, severe and crippling lameness, reproduction problems including spontaneous abortions, and even death. Not surprisingly, milk from affected cows is likely to contain more antibiotics, hormones, and dispersed pus; yet the Food and Drug Administration allows such milk to be sold without labeling that identifies its source.

Reprinted by permission of Kirk Anderson.

Until now the genetic engineering of animals has most often involved either the use of a genetically engineered hormone on animals or the transfer of a single gene between species. Most observers believe, however, that in the next few years scientists will be able to manipulate animals genetically in far more ambitious ways. Recently, the congressional Office of Technology Assessment (OTA) predicted that by 2005 scientists will be engineering complex genetic traits, including those involving human behavior, into other species. Researcher J. Mintz, who successfully transplanted rabbit growth genes into mice, predicts that soon we may see "five-ton cows and pigs twelve feet long and five feet tall." Other researchers have predicted the creation of a wide variety of transgenic creatures, including monster chickens, oysters that survive in polluted waters, and wolves that avoid sheep. "Right now, we don't know what the limits are," said Michael Phillips of the OTA. "All the traditional rules we thought about the . . . animal kingdom . . . are thrown out the window."

Some genetic engineers are more conservative about prospects for the future. "We're at the Wright Brother stage compared to the 747," says Pursel. "We're going to crash and burn for a while." Of course, it is animals like Pursel's pig No. 6707 that pay the price for researchers' "crash and burn" persistence.

Leaner, more abundant meat is not the only goal of genetic engineers. Some are attempting to create valuable research subjects by engineering genes for cancer and other human diseases and disorders into the permanent genetic codes of animals. Biotechnologists hope that these animals will be superior research models on which to test important new drugs and therapies.

The creation of transgenic animals containing human disease genes, however, poses unique environmental, human-health, and animal-suffering hazards. A remarkable and timely example is the AIDS mouse.

Normal mice do not make good AIDS-research tools because they are not susceptible to infection with AIDS. Thus, National Institutes of Health (NIH) researcher Malcolm Martin transplanted the entire AIDS virus genome into numerous mouse embryos in the fall of 1987. Ten percent of those mice carried the AIDS virus in their permanent genetic make-up at birth.

Potential for Disaster

The creation of the AIDS mice was hailed as a breakthrough by the media and by many in the scientific community. Other scientists and observers, deeply troubled by the experiments, worried that Martin, by creating a nonhuman repository for a human disease as volatile and as fatal as AIDS, had created a potential nightmare of extraordinary proportions. If the genetically engineered animals escaped and mated with other mice, there could be an overwhelming and uncontrollable risk of spreading AIDS.

Martin dismissed such fears as "far-fetched" and a *National Enquirer* scenario," but a year after his experiment had started, disaster struck. According to NIH sources, on Saturday, December 3, 1988, electric power providing air to the AIDS mice was apparently shut off inadvertently by a repair man. When researchers came in early Sunday morning, they found the laboratory without power, and only 3 of the 130 AIDS mice were alive. Martin did not explain how this kind of accident could occur in an experiment that was, purportedly, among the most carefully monitored in NIH history.

An even greater problem lay ahead for Martin. In February 1990 a paper that appeared in *Science* reported on experiments conducted by a distinguished team of scientists headed by Robert Gallo, the codiscoverer of the AIDS virus. The Gallo team reported that HIV could interact and combine with the natural viruses of transgenic mice used in AIDS research. This interaction could produce new and potentially more dangerous forms of HIV—a super AIDS that might infect a wider range of cells and be spread by novel routes. According to AIDS researcher Jean Marx, one of these novel routes might be airborne transmission of the new super-AIDS virus.

A Disorderly Retreat

The findings of the Gallo team mirrored and vindicated many of the scientific and safety concerns of those who had been critical of Martin's experiments. The idea that inserting the AIDS virus into the permanent genetic code of mice or other species might enhance the virus' virulence was not a far-fetched tabloid scenario, but a scientific probability. The theory that transgenic animals could be made into perfect, or even useful, research tools for HIV had been undermined seriously. Like Pursel's, Martin's highly publicized march into the transgenic future had become a disorderly retreat leaving animal suffering, media hype, false hopes, and large taxpayer expenditures in its wake.

In addition to the tens of thousands of animals genetically engineered to make better food sources and research tools, other animals are being altered to produce medically valuable human biological materials. This line of research began in the early 1980s when scientists engineered sheep that created human growth hormone in their blood and pigs that created human insulin in a like manner. After human genes that had become part of the animal's permanent genetic codes had created sufficient amounts of the desired human product, the animal was slaughtered, and the valuable human hormone or other biochemical was harvested. Because slaughter had the obvious disadvantage of destroying the animal prior to retrieving the product, researchers began looking for ways to engineer animals to produce valuable human materials on a sustained basis and to harvest those chemicals without having to destroy the new animal "factories."

Biochemicals and Milk

At last, researchers think they have found a viable solution: genetically engineering new breeds of sheep, goats, and cows that secrete valuable human pharmaceuticals in their milk. There was early skepticism about this approach, but in 1987, researchers were successful in engineering a human gene into mice that made them produce a human protein in their milk. In September 1991 three independent research teams jolted the biotechnology world by announcing major advances in getting animals to secrete valuable biochemicals in their milk.

"Two years ago people were doubtful of this technology," said animal biotechnologist Robert Bremel of the University of Wisconsin, "but now the work shows that the mammary gland can be used as an impressive bioreactor."

From the first, critics have raised fundamental ethical and moral concerns about humankind's right to use genetic engineering to redirect the evolutionary process of animals in the name of profit and efficiency. Critics have noted that genetic engineering not only causes great animal suffering but also puts the genetic integrity of many species of animals at risk. Unfortunately, too many scientists and corporations and the federal government continue to ignore the animal suffering and the ethical questions that surround the genetic engineering of animals.

"*Studies have clearly shown the efficacy, safety and benefits realized by integrating bST into dairy production.*"

Adding Growth Hormones to Dairy Cattle Is Beneficial

Terry D. Etherton

Genetic engineering methods are being used to alter farm animals so that they can produce more milk and eggs and better quality meat. One such method is the injection of a hormone, bovine somatotropin (bST or BST), into dairy cows to increase their milk production. In the following viewpoint, Terry D. Etherton argues that BST is a safe, healthy, and effective way to increase milk production. He believes that increasing the productivity of dairy cattle and other farm animals is important for the world to be able to feed its growing population. Etherton is a professor of animal nutrition and physiology in the department of dairy and animal science at Pennsylvania State University in University Park.

As you read, consider the following questions:

1. How does BST affect the composition of milk, according to Etherton?
2. Critics of BST contend that it increases infections in cattle. How does the author refute this claim?
3. What statistics does Etherton cite to show that BST is safe for humans?

Excerpted from *The Efficacy, Safety, and Benefits of Bovine Somatotropin and Porcine Somatotropin* by Terry D. Etherton. Copyright 1994 by the American Council on Science and Health, Inc. Reprinted with permission.

On November 5, 1993, the Food and Drug Administration (FDA) approved the first biotechnological product for animal production, bovine somatotropin (bST), for commercial use. This action ushered in a remarkable new era for animal agriculture and the dairy industry. BST use results in a substantial increase in milk yield (4 to 6 kg/day or 10 to 15%) accompanied by an approximate 12% increase in productive efficiency. Milk yield increases in a dose-dependent manner and the composition of milk is unaltered.

Scientists in academia, government and industry have conducted more than 2,000 scientific studies of bST throughout the world. These studies have clearly shown the efficacy, safety and benefits realized by integrating bST into dairy production. BST does not adversely affect the health of treated cows. Supplemental administration of bST does not affect the quantity of bST found in milk or the milk's composition. Milk and meat derived from bST treated cows are safe for human consumption.

Hormones for Pigs

A second biotechnology product, porcine somatotropin (pST), likely will be approved shortly by the FDA. When maximally effective doses of pST are given to growing pigs, there is a marked increase in growth rate (10 to 15%), productive efficiency (body weight gain/feed consumed) is increased by as much as 15 to 35%, and carcass fat is reduced by as much as 80%. Producing leaner pork will benefit consumers who wish to reduce their intake of total and saturated fat.

Technologies that lower the quantity of feed consumed per unit of output (meat or milk) will also benefit the producer—because feed constitutes the major component (about 50 to 70%) of animal production costs. The use of bST and pST will also have a beneficial effect on the environment. A reduction in the amount of feed required to produce a unit of meat or milk would reduce the need for fertilizer and other inputs associated with growing, harvesting, processing and storing animal feed. Increases in productive efficiency reduce the production of animal wastes including methane.

Although the biotechnological advances made in the past 15 years have been accepted by the scientific community, public misunderstanding about the safety and benefits of bST and pST endures. Many consumers do not receive factual information or, even worse, hear misinformation about biotechnology. Thus, it is difficult for them to evaluate accurately the safety and benefits of these products. It is important that credible scientists and scientific organizations make an effort to inform consumers about the benefits and safety of bST and pST. To benefit society, biotechnology products must provide tangible benefits and be accepted by the public as safe.

Somatotropin (ST) is a naturally occurring protein hormone produced in and secreted from the anterior pituitary gland. It is the master hormone regulating the growth of humans and other animals. It also markedly stimulates milk production and improves productive efficiency.

Insufficient production of ST during childhood markedly retards human growth, resulting in short stature at maturity (dwarfism). In advanced countries, dwarfism is not a common occurrence because physicians can diagnose and treat it with recombinantly derived human ST (hST). Because of the remarkable effects that ST has on growth, it also is referred to as growth hormone. However, because somatotropin has many important and diverse biological roles that affect the metabolism of all classes of nutrients, it is preferable to call it somatotropin.

All protein hormones, including ST, are made from naturally occurring amino acids that are derived from the digestion of dietary protein. BST contains 191 amino acids and shares a high degree of amino acid sequence similarity in the range of 90% with somatotropins from other farm animals. Although the amino acid sequence is quite similar among farm animals, there are unique differences when comparisons are made with human ST. For example, the amino acid sequence of hST differs by approximately 35% from that of bST. Because of this, bST is not active in humans even if it is injected into the bloodstream. This is an important attribute assuring its safety for the consumer.

What Is a Hormone?

The circulatory systems of humans and animals are packed with chemical messengers called hormones. These hormones travel through the bloodstream in search of specific hormone receptors on distant cells. There they bind to these receptors and initiate a vital cascade of life-sustaining events inside each cell. Hormone receptors are extraordinarily selective in their ability to recognize and bind to hormones. This property confers great specificity in hormone action.

The behavior of both humans and animals is governed by hormonal signals which are received, decoded and acted on by the appropriate cells that make up tissues and organs. A cell that can respond to a specific hormone is called a target cell for that particular hormone. There is great specificity in the endocrine system. Not all cells respond to all hormones; thus, certain hormones may have very potent effects in some cells and no effect in others. . . .

Bovine somatotropin does not change the composition of milk in any significant way (Table 1). The concentration of fat and protein in milk varies due to genetics, stage of lactation, age, diet composition, nutritional status, environment and season. These factors also affect the composition of milk from bST-supplemented cows. Any minor differences in milk composition from bST supplementation are within the normal range. The variations in the content of fat and protein in milk are of the same magnitude as those usually observed in cows not supplemented with bST.

Table 1. Effects of bST on Milk Composition

	from bST-supplemented cow	from control cow
% Fat	3.7	3.8
% Protein	3.2	3.3
% Lactose	4.8	4.9

Terry D. Etherton, *The Efficacy, Safety and Benefits of Bovine Somatotropin and Porcine Somatotropin*, July 1994.

Many detailed chemical analyses of milk from bST-treated cows have been conducted. Generally, bST supplementation does not alter the proportion of total milk protein represented by whey proteins and caseins. Milk from cows supplemented with bST does not differ in the quantity of vitamin A, thiamin, riboflavin, pyridoxine, vitamin B-12, pantothenic acid or choline; the content of biotin increases slightly. . . .

Studies have also determined that bST does not affect milk flavor nor any manufacturing characteristics that are important during the production of processed dairy foods such as cheese or yogurt.

The nutritional needs and responses to adjusting the diets of cows supplemented with bST have been reviewed in depth. Studies have examined the production responses to bST under a wide variety of feeding programs. Obtaining a response in milk production to bST does not require special diets or unique feed ingredients. It is important that the diet meet the cows' nutrient requirements, which are influenced by the milk yield. Cows supplemented with bST increase their food intake to provide the extra nutrients needed to sustain the increased milk production, but the nutrient composition and density of the diet do not need to be modified.

Cows typically adjust their voluntary feed intake upward within a few weeks after initiation of bST supplementation. Thus, to maximize the milk response to bST, dairy farmers must be attentive to management factors that affect food intake. High quality forage is a critical component in obtaining high levels of voluntary intake. Other important factors that farmers must consider to optimize the response to bST are: ad libitum feeding (free access to feed at all times), unlimited access to clean cool water, nutritionally balanced diet, adequate dietary protein, proper levels of digestible fiber and control of temperature and humidity. If cows consume an insufficient quantity or imbalanced composition of nutrients, the response to bST will decrease according to the extent of the inadequacy.

Animal Safety and Health

In addition to establishing the effectiveness of bST, companies seeking regulatory approval of the product were required by FDA to prove that it was safe. Evidence of bST's safety was required in three areas: 1) the safety of animal food products (milk/beef) for humans, 2) the safety of bST for the treated cow and 3) the safety of bST for the environment.

Safety concerns were seized upon by some opponents of biotechnology in an attempt to misinform the public and politicians about bST. They waged their campaigns in an attempt to gain support for their cause and to hinder FDA approval of the product. Opponents made unsubstantiated claims in the popular press that bST would reduce resistance to infectious diseases and thus increase the incidence of sickness in dairy cows. Responsible scientists and scientific and governmental organizations addressed this issue rigorously and concluded that bST is safe and does not adversely affect animal health.

To verify that bST is safe for dairy cows, the FDA required that safety margins be established. This was done by treating cows with 60 times the commercial dose of bST over a two-week period and up to 6 times the commercial dose for two consecutive lactations. These studies showed that administration of large doses of bST did not affect animal health. A recent review of the literature reinforces this important point. D.E. Bauman surveyed hundreds of studies in which cows were supplemented with bST and did not find a single study that linked bST with an increased incidence of ill health.

There have been some claims in the popular press that bST increases the incidence of clinical mastitis (inflammation of the udder). Susceptibility to mastitis is related to many factors including, but not limited to, season, stage of lactation, environment and milking management. A recent report evaluating 15 full lactation trials (914 cows) and 70 short-term studies (2,697 cows), after accounting for all major factors, found that a small positive association remained between milk yield and mastitis on a per cow basis. This relationship allows critics to make the increased mastitis charge, but their interpretation is erroneous.

BST supplementation causes no more mastitis than would have occurred with any increase in milk yield. More importantly for farmers and consumers, mastitis incidence declined slightly per volume of milk produced. In addition, there is evidence which indicates that bST supplementation is associated with a more rapid recovery of milk secretion in mammary glands of cows experimentally infected with bacteria that cause mastitis. A recent publication by the International Dairy Federation confirmed that treatment with bST has no effect on the incidence of mastitis.

BST Is Safe for Humans

BST is naturally present in cows' milk and meat. With respect to meat, bST supplementation of cows for 28 days does not significantly increase bST concentration. Even if there were an increase in ST concentration in meat as the result of ST supplementation, this would not be a human food safety concern. The scientific basis for the FDA conclusion that milk and meat from bST-supplemented cows is safe is based on the following scientifically established points:

1. BST is a protein and, like all other plant and animal proteins in the diet, it is digested to single amino acids and very small peptides through a combination of the low pH conditions and the digestive enzymes found in the stomach and upper small intestine. The resulting amino acids have no hormonal activity. This has been confirmed in studies reviewed by the FDA in which bST was administered orally to rats for 90 days at doses up to 50 mg/kg/day. Each study met the FDA's minimum requirement of at least 14 days of supplementation with up to 100 times or more of the bST dose projected to be used commercially for the target animal (based on a mg/kg body weight basis).

2. In addition, pituitary-derived somatotropin from farm animals and other nonprimates is inactive in human children and adults even when injected. Subsequent research has shown that this "species-limited" effect occurs because the amino acid sequence of bST differs from that of hST by approximately 35%. As a result, somatotropin receptors in human cells cannot recognize bST. Thus, even if a human were injected accidentally with bST, there would be no hormonal effect.

"Major questions about BGH's impact on human and animal health remain unanswered."

Adding Growth Hormones to Dairy Cattle Is Harmful

Tim Atwater

Bovine growth hormone (known as BGH or BST), which is injected into dairy cows to increase their milk production, was approved for commercial use by the Food and Drug Administration (FDA) in February 1994. In the following viewpoint, which was written prior to the FDA approval, Tim Atwater presents several common arguments against the use of the hormone. He contends that the United States is currently experiencing a milk surplus, and that increasing milk production will harm farmers economically. In addition, according to the author, the hormone poses potential health threats to both animals and humans. Atwater is a reporter for *Catholic Worker*, a bimonthly newspaper of religious and social issues.

As you read, consider the following questions:

1. How do consumers feel about the addition of growth hormone into milk, according to the poll cited by the author?
2. How will BST affect genetic diversity, in Atwater's opinion?
3. What final comment do the editors of *Catholic Worker* make concerning the use of antibiotics in farm animals?

Tim Atwater, "Unnatural Growth Hormone," *The Catholic Worker*, March/April 1993. Reprinted with permission.

Agribusiness corporations are pushing a new technology to make cows produce more milk. Bovine growth hormone (BGH or BST) is a synthetic, genetically engineered version of a cow's naturally occurring growth hormone. Cows injected with BGH in test herds produce an average of 10% to 25% more milk. The federal Food and Drug Administration (FDA) is considering applications to permit marketing of BGH submitted by four drug and chemical companies. [Final approval to market BGH was granted in February 1994.]

One problem is that we don't need more milk. Even a slight increase in milk production can lead to a drastic decline in milk prices. In 1990–91, a 3% increase in milk production led to a 35% decline in dairy prices.

And consumers don't want hormones in their milk. A 1990 poll by the National Dairy Promotion and Research Board (a politically appointed body which has been bending over backwards to help industry promote BGH as "safe and natural") shows that between 41% and 50% of consumers nationally would curb or halt their purchases of dairy products if they thought cows were treated with BGH—even when BGH was described in very positive terms by pollsters.

Treating Sick Cows

Data from the University of Vermont/Monsanto BGH test herd shows that cows injected with BGH experienced a marked increase in animal health problems. This means more veterinary bills, extra expenses for replacement cows, and consumer concerns about possible excessive use of antibiotics to treat sick cows.

BGH has not yet been approved for commercial use by the FDA. Widespread testing of BGH has been approved, however, and the FDA has allowed the sale of milk and beef from test herd dairy animals since 1985. If approved, BGH could be one of the first genetically engineered farm and food products to be widely available in the marketplace.

Monsanto, Eli Lilly, American Cyanamid, and Upjohn have over $500 million invested in BGH to date, with about half of it from Monsanto alone. These companies, hoping for annual sales of $400 million or more, have found willing ears within the FDA, and only pressure from citizen action groups on members of Congress has kept the FDA from granting approval.

BGH has profound implications for the future of dairy farming, animal genetic diversity, and the development of commercial agricultural biotechnology. And major questions about BGH's impact on human and animal health remain unanswered—

Family Farms. BGH threatens the future of dairy farming by undercutting consumer confidence in the safety of dairy products and by creating price-depressing milk surpluses. Dairy industry studies have projected a loss of up to 50% of the nation's dairy farms from BGH.

Human Health. Consumers are justifiably wary. In recent years federal regulators have had poor records in identifying health and safety risks associated with chemicals, food additives and other threats to consumer health. A 1990 Congressional study found that more than half of the drugs approved by FDA over a ten-year period were later found to have "serious post-approval risks" for consumers. An August 1992 report by the Congressional General Accounting Office says BGH increases the incidence of mastitis, an udder disease usually treated with antibiotic drugs, and concludes BGH should not be approved unless further testing proves that the drug doesn't increase antibiotic use.

An Emotional Issue

The fact that someone is "doing something" to milk is clearly a deeply emotional issue. Even people who try to avoid fat and cholesterol feel protective about milk, because of its wholesomeness and because children drink it to grow strong. Perhaps we can't help having such feelings—as mammals we may have a built-in reverence for milk. In the month that BGH was approved for public consumption, milk sales across the country dropped dramatically. Night after night on the network news, people identified as "concerned customer" or "former milk buyer" complained that "It isn't natural" and "I don't know what it is, but I don't want it in my kids' milk."

Tony Hiss, *Harper's Magazine*, October 1994.

Animal Health. A 1991 report by the Rural Vermont Farm Advocacy Group, based on previously secret data from the University of Vermont/Monsanto test herd, shows that cows injected with BGH become ill, give birth to dead and deformed calves, and are sold for beef because of shortened milking lives far more frequently than other cows. Data suggests that some daughters of BGH cows (who were not themselves injected with BGH) gave birth to seriously deformed calves, raising the possibility that genetic damage could have occurred among BGH-treated cows.

Breeding Records Affected

Loss of Genetic Diversity. BGH represents a major step toward the loss of genetic diversity in dairy animals. If BGH is commercially marketed we will see increased use of breeds of cows that react most positively to the hormone treatment. Jerseys and some other breeds seem to have significantly higher negative reactions to the hormone. Breeding records will be seriously affected by BGH, since it could be impossible to determine a cow's genetic "value" accurately if some or all of its offspring have been treated with BGH. Genetic diversity is likely to suffer.

Unfortunately, BGH is only the tip of the iceberg. Agribusiness is already well along in the development of patented "designer" genetically engineered cows—among the many other farm and food products scheduled to be produced through biotechnology. Here is an issue that cries out for wisdom and resistance—particularly from the churches.

The Degradation of Antibiotics

Note from editors of *Catholic Worker:* Antibiotics, which are indiscriminately added to animal feeds or injected into livestock in order to increase short-term productivity and profits, are later absorbed by the human beings who consume the meat. This promiscuous use of antibiotics promotes the development, in nature, through adaptive mutations, of new, drug-resistant strains of micro-organisms. Thus, for example, penicillin, the first antibiotic, discovered in 1929 by Alexander Fleming, is no longer as effective today as it once was, and new generations of antibiotics are having to be developed at great expense and time. The agribusiness proposal to use antibiotics indiscriminately in animals will accelerate the degradation of the potency and efficacy of current antibiotics, on which we depend when stricken with bacteria-generated diseases.

Periodical Bibliography

The following articles have been selected to supplement the diverse views presented in this chapter.

The Amicus Journal	"The Splice of Life," special section on biotechnology and ecology, Spring 1993.
Cheryl Cook	"When Best Isn't Good Enough," *Christian Social Action*, January 1991. Available from 100 Maryland Ave. NE, Washington, DC 20002.
R. James Cook	"Biotechnology Applied to Microbes: Taking Agricultural Sustainability a Step Forward," *The World & I*, December 1994. Available from 2800 New York Ave. NE, Washington, DC 20002.
Norman Ellstrand	"How Ya Gonna Keep Transgenes Down on the Farm?" *The Amicus Journal*, Spring 1993.
Michael W. Fox	"Genetic Engineering and Animal Welfare," *Applied Animal Behavior Science*, vol. 22, 1989. Available from PO Box 211, 1000 AE Amsterdam, Netherlands.
Carol Grunewald	"Monsters of the Brave New World," *New Internationalist*, January 1991.
Paul Hatchwell	"Opening Pandora's Box: The Risks of Genetically Engineered Organisms," *The Ecologist*, 1989.
Sandy Miller Hays	"Farm Animals of the Future," *Agricultural Research*, April 1989 and May 1989. Available from the Government Printing Office, 732 N. Capitol St. NW, Washington, DC 20401.
Richard Hindmarsh	"The Flawed 'Sustainable' Promise of Genetic Engineering," *The Ecologist*, September/October 1991.
Margaret Mellon	"Engineering the Environment," *Christian Social Action*, January 1991.
Bernard E. Rollin	"The Frankenstein Thing: Ethical Issues in Genetic Engineering," *USA Today*, November 1990.
The World & I	"The Ethics of Biotechnology," special section, December 1994.

Is DNA Fingerprinting Accurate?

**GENETIC
ENGINEERING**

Chapter Preface

Police forensic experts have long used the fingerprints left at crime scenes to identify suspects. Such fingerprints are still considered by investigators to be the most accurate crime-scene evidence. But investigators now have another investigative tool at their disposal: DNA fingerprinting.

DNA fingerprinting is a process by which people can be identified based upon the DNA patterns found in hair, saliva, blood, and other body tissues. It has been used in numerous cases to both convict and clear criminal suspects. For example, investigators linked the DNA in the saliva used to seal an envelope to one of the suspects in the 1993 bombing of the World Trade Center. On the other hand, according to journalist William Tucker, "30 percent of [DNA] tests yield negative matches, exonerating innocent suspects."

Many forensic experts view DNA as the ultimate tool in solving crimes. "DNA-based identity testing is probably the most significant development in forensic science since fingerprinting itself, creating the possibility of uniquely identifying an individual from a single cell left at the scene of a crime," writer Howard Cooke states in the British journal *Lancet*. But the technology is not without its critics, many of whom question the accuracy and competence of the labs that perform DNA testing. As Peter Doskoch writes in *Science World*, "DNA fingerprinting may sound like the ideal evidence for an open-and-shut case—if the tests are done accurately. But that, say opponents of using DNA evidence, is a big 'if.'"

The contributors to the following chapter examine the potential benefits and harms of DNA fingerprinting and discuss how this new technology might affect criminal justice.

"DNA profiling has implicated the guilty and exonerated the innocent in a way that was previously unthinkable."

DNA Fingerprinting Is Reliable and Accurate

William Tucker

The following viewpoint was written as former professional football player O.J. Simpson went on trial for the murder of his former wife and her friend. Anticipating the role of DNA evidence in the trial, freelance writer William Tucker describes the development of the technique of "DNA profiling" and counters the arguments against its use in courts. He contends that DNA fingerprinting is a reliable and accurate criminal justice tool that can assist in both identifying the guilty and exonerating the innocent. Tucker is the New York correspondent for the *American Spectator*, a monthly news and opinion magazine.

As you read, consider the following questions:

1. How does the certainty of DNA profiling compare with that of other forensic evidence, according to Tucker?
2. According to the author, why is it misleading to compare DNA profiling to the Greiss test?
3. Why are "ethnic ceilings" on DNA profiles largely irrelevant, in the author's opinion?

At the time of this writing, November 1994, O.J. Simpson is standing trial for murder in Los Angeles in what could be one of the most gripping courtroom dramas in decades. Accused of killing his former wife, Nicole Brown Simpson, and a young visitor, Ronald Goldman, Simpson has persistently denied his guilt. . . .

The case against the former football great is largely circumstantial, and there are weaknesses to the case that might make an American jury throw up its hands and declare reasonable doubt. There are no eyewitnesses to the crime. No video cameras were on hand to record the event. The murder weapon has never been found. The policeman who found a bloodstained glove at Simpson's house (matching another found at the crime scene) has a history of making disparaging remarks about minorities.

Powerful Evidence

There is one piece of evidence, however, that is so powerful that it might easily erase any reasonable element of doubt from the jury's mind. Leading away from the crime scene was a trail of fresh blood, which apparently belonged to the murderer. Simpson had a cut on his left index finger when questioned by police the next day. At least two separate laboratories have examined this blood, using the technique of "DNA profiling," and have declared that it matches Simpson's DNA profile.

Yet, as the trial began, there was a serious question of whether the jury would be allowed to hear this evidence. [The DNA evidence was subsequently allowed.] Moreover, if Simpson is declared guilty, his conviction will be challenged in a long series of appeals. A California appellate court has already overturned one conviction in a similar case, arguing that DNA evidence was "unreliable." The same situation prevails in eight other states.

How did DNA profiling—almost a decade old and widely employed in other countries—end up having such a rough ride through the American justice system? The science itself is not at issue. There has never been a case where one laboratory declared a match in DNA samples and another laboratory declared the opposite. Believe it or not, the only major controversy now surrounding the technique is whether the chances of an innocent person being falsely implicated are 1-in-10,000 (a high estimate arbitrarily chosen by a maverick scientist) or 1-in-10 million (a widely accepted figure that has been verified by an examination of all the DNA records on file with the FBI).

Other forensic evidence long accepted in American courtrooms offers levels of certainty that are nowhere near that range. Blood-type identification, accepted in courts for decades, offers at best only a 90 percent verification (1-in-10 possibility of a chance match-up). Handwriting analysis and psychiatric

122

testimony in insanity cases usually come down to a "battle of experts." Only with "dermatoglyphic" fingerprinting (the marks on the end of your finger) are the probabilities of the same general order of magnitude. Yet with DNA profiling, defense attorneys have successfully argued that, if scientists cannot agree whether the technology is 99.99999 percent certain or 99.99999999 percent certain, then *it shouldn't be used at all*.

Junk DNA

DNA profiling begins with the established theory that no two people, except identical twins, have the same genetic makeup. Each cell in the body contains a complete set of genes. A clot of blood, a trace of skin underneath a victim's fingernails, a drop of semen, the follicle attached to a single strand of hair—all contain enough cells to provide the information for a positive or negative match with a criminal suspect. . . .

In 1985, Alec Jeffreys, a geneticist at the University of Leicester, England, proposed making forensic identifications with "junk" DNA, the mysterious, non-functioning genetic material that makes up about 95 percent of the human genome. This material serves no known purpose. It may just be "hitchhiking" from generation to generation without contributing anything to the organism. Or it may serve as "packing material," protecting the working genes from harmful mutations, the way newspapers stuffed in a box will protect its fragile contents.

Junk DNA varies from one individual to the next. Different people have different DNA sequences at their junk sites. In addition, these characteristic sequences repeat themselves a different number of times in different people—a phenomenon called "variable number tandem repeats" (VNTRs). One person may have only one repetition at his junk site, while another may have two dozen. Most sites have more than a hundred known variations, which are called "alleles."

Other genetic markers such as hair color, height, and weight tend to vary by population. People living near the equator, for example, generally have darker skin, while people in cold climates generally grow bulkier to conserve heat. VNTRs, however—like fingerprints and blood types—appear to vary randomly across populations, with no ethnic or racial associations.

Criminal Identifications

In 1986, Jeffreys proposed that VNTRs could be used for criminal identifications. He invented a "multi-locus" molecular probe that surveyed about fifteen to twenty VNTR sites, measuring their varying lengths. The chances that any two people would have the same variation at one site is about 1-in-50. The chance that they would match up at *every* one of the fifteen to twenty sites is well beyond 1-in-1 trillion. (The whole earth's population is only 8 billion.)

The test is now used in paternity suits. In criminal cases, however, "multi-locus" probes did not always prove practical. "The difficulty is that we rarely have enough genetic material in the sample," says Mark Stolorow, director of operations at Cellmark Diagnostics, which is running the tests in the Simpson case. "With paternity suits, we can just take blood samples out of someone's arm. But in criminal cases, we're often dealing with a speck of blood found on the sidewalk." Thus, Jeffreys's multi-locus "genetic fingerprinting" (the name is trademarked) was supplanted by a "single-locus" probe, which, given about 8,000 cells (the amount in a drop of blood), can provide a "genetic profile" with somewhat lower degrees of certainty.

Protecting the Innocent

There has never been a case in which a person has been convicted using DNA evidence and later been proven to be innocent. In fact, people have actually been freed from prison thanks to DNA evidence that turned up after they were convicted.

Rockne Harmon, quoted in *Time*, August 8, 1994.

In 1987, Jeffreys licensed his technology to Imperial Chemical, a British firm, which set up Cellmark Diagnostics, in Bethesda, Maryland. Lifecodes, Inc., now in Stamford, Connecticut, also went into the business, using a slightly different technology. Eighty different state crime labs, plus the FBI, have also entered the field. About 4,000 samples of DNA were tested last year, at an average of $1,000 per test. The number of probes used depends on how much genetic material is available and how much a prosecutor wants to spend. At five probes, the theoretical chances of two individuals having the same profile are $1\text{-in-}50^5$, or 1-in-312 million.

In actuality, the alleles do not occur with the same frequency. Some are common while others are rare. If you have common alleles, you may match with 2,500 other people in the country (1-in-100,000), while if your alleles are rare, the match may be only 1-in-1 billion. In 1992, Neil J. Risch and Bernard Devlin of Yale University, using the FBI's database, generated 7.6 million genetic fingerprints and found only one chance match at the *three*-probe level. At the four-probe level there were none. They estimated the chances of a match for five probes at 1-in-10 billion. . . .

From its inception, DNA profiling has implicated the guilty

and exonerated the innocent in a way that was previously un-thinkable. In an early case in England, two adolescent girls in a small village had been raped and murdered over a three-year period. Police asked males in the village to give a DNA sample for comparison. No matches were found, but it was later re-ported that one Colin Pitchfork had bribed someone else to sub-stitute a sample for him. Pitchfork was checked again and turned out to be a match. (This case was chronicled by Joseph Wambaugh in *The Blooding*.)

In an early incident in the United States, a young couple were murdered at an isolated campground in Colorado. The woman had been raped and a semen deposit was found. A random check against profiles of known sex criminals turned up a match with a paroled felon in Florida. Once he was under suspicion, eyewitnesses were able to place him near the scene of the crime. The man was tried and convicted. . . .

DNA profiling has proved just as important in clearing the in-nocent as it has in implicating the guilty. American laboratories report that 30 percent of tests yield negative matches, exonerat-ing innocent suspects who would otherwise have gone to trial. Scotland Yard [police headquarters in London, England] reports the same percentages.

The Counterattack

So things stood until 1989, when a handful of lawyers mounted a counterattack. The principal players have been Peter Neufeld, a New York defense attorney, and Barry Scheck, a professor at the Benjamin Cardozo School of Law in New York. "The attitude up to that point had been that DNA fingerprinting was infalli-ble," said Neufeld. "Juries were awed. As one juror put it, 'You can't argue with science.' We decided to show you could." Neufeld has not only carried through the battle in court, he has also succeeded in becoming the resident expert on the subject in the pages of *Scientific American*. . . .

The first important case involved Jose Castro, a South Bronx janitor accused of stabbing to death Vilma Ponce and her two-year-old daughter in 1987. When Castro came under suspicion, a speck of blood was found on his watch. The sample was sent to Lifecodes, which said it belonged to the victim. Neufeld and Scheck challenged the admissibility of the evidence on the grounds that the lab work was sloppy and there were too many uncertainties in the technology.

Genetic experts from both sides converged on the scene. Before testimony began, Eric S. Lander, of Massachusetts Institute of Technology's Whitehead Institute for Biomedical Research, testi-fying for the defense, and Richard J. Roberts, of Cold Spring Harbor Laboratories, testifying for the prosecution, decided to get together and issue a joint statement. Both were somewhat disenchanted with Lifecodes's performance.

In particular, they were concerned that Lifecodes was declaring matches in instances where the X-ray images that read the VNTRs were identical but shifted slightly out of place—a phenomenon called "band-shifting." The laboratories claim it is not a problem. "It's like having two pieces of identical wallpaper that are hung poorly," says Michael Baird, lab director at Lifecodes. "You can see the patterns are identical, but they're slightly displaced."

Lander and Roberts argued that band-shifting created too much uncertainty. They also pointed out that Lifecodes had declared one match when the bands were outside the 5 percent range of error. In a blind test submitted by the California Association of Criminal Laboratory Directors, Cellmark had also misread one sample in fifty as a match. In 1989, Judge Gerald Sheindlin threw out the evidence tying the blood of the victim to Castro's watch—although evidence showing Castro himself was not the source of the blood was admitted. Castro pleaded guilty anyway and was sentenced to a lengthy prison term. . . .

A Misleading Analogy

In 1990, in *Scientific American*, Neufeld laid out the full case against DNA fingerprinting. Neufeld compared DNA profiling to the Greiss test, a chemical test for nitrates from explosives, which had been used to convict six Irishmen in an IRA [Irish Republican Army] bombing. "It turns out that a variety of common substances such as old playing cards, cigarette packages, lacquer and aerosol spray will, along with explosives, yield a positive result [in the Greiss test]," wrote Neufeld. Neufeld then outlined similar potential flaws in DNA profiling: samples were small, DNA could be changed by the presence of impurities and bacteria, the sample might degenerate in a number of ways. The band-shifting problem distorted results. Samples could be accidentally switched or mislabeled—any number of things might happen. As a result of all this an innocent person might be convicted of a crime.

But Neufeld's opening analogy was misleading. The major problem with the Greiss tests was that it produced false positives. Substances other than the target chemical could give the same results. With DNA analysis, however—and particularly with the problems mentioned by Neufeld—the only real problem is false *negatives*. The chances of an innocent person being implicated are next to nil, but the chances of a guilty person being falsely exonerated are reasonably high.

To simplify, suppose that a suspect has a five-allele code that reads: 26-13-12-27-11. The forensic sample, which also contains his genes, has the same code. Now suppose the forensic sample degenerates, as Neufeld suggested. It can only degenerate *away* from a positive match. (In practice, the lab would probably call the results "inconclusive," which happens in 10 to 30 percent of all tests.)

Now suppose the suspect is innocent. What are the chances that a forensic sample will degenerate *into* his code of 26-13-12-27-11? They are, in fact, approximately the same as the likelihood that a chance mismatch will occur in the first place—about 1-in-10 million.

The great irony is that, while arguing that DNA profiling should not be used *against* criminal suspects, Neufeld and Scheck are simultaneously representing 600 condemned prisoners who claim that DNA analysis will prove they are innocent. Despite the much greater problem of false negatives, the attorneys argue that DNA evidence is valid when used on the side of the defense.

DNA and Ethnic Groups

As a final argument against admissibility, Neufeld also raised what was soon to become the principal objection to DNA profiling: the idea that the genetic markers used in DNA analysis are not randomly distributed by racial groups, that they follow the pattern of hair and eye color, rather than blood types and fingerprints. Thus, when compared against people in a suspect's own racial or ethnic group, the chances of an accidental match-up might be higher.

The argument was later expanded by Richard Lewontin, a maverick population geneticist at Harvard and co-founder (with fellow Harvardian Stephen Jay Gould) of the left-wing academic group Science for the People. In 1991, Lewontin co-authored an article in *Science* that argued that patterns at separate VNTR sites might be inherited as a unit, creating similar genetic profiles among small, inbred populations. This "pose[s] a particularly difficult problem for the forensic use of VNTRs if the wrong ethnic group is used as the reference population." In order to avoid chance mistakes, it would be necessary to develop much more data about "subgroups that are likely to be relevant in forensic applications." The authors identified these groups as blacks, Hispanics, and Amerinds, and speculated that the chances of a false match-up within these populations might be as high as 1-in-10,000.

Now, 10,000-to-1 is still pretty long odds—certainly enough to erase any element of reasonable doubt where other incriminating evidence is present. But Neufeld wanted to go a step further. Instead of merely increasing the odds, he now argued that there was no "consensus" about DNA technology in the scientific community and therefore the technique should be excluded altogether from criminal trials. Appeals courts in California, Massachusetts, Arizona, Minnesota, and five other states bought the argument and previous convictions were overturned in each state.

The unsubtle point behind Lewontin's talk of forensic "relevance" was this: since blacks, Hispanics, and Indians commit a disproportionate share of all crimes, an individual *within* one of these groups may end up being implicated by the newfangled technology. (Actually, the black population has proved to be more genetically diversified than any other racial group.) As a later *Scientific American* article put it, "An innocent suspect racially or ethnically similar to that of a criminal could have an inflated chance of matching a forensic sample—and thus be wrongly convicted."

All this assumes that suspects are implicated in crimes solely on the basis of their race—which in some cases they are. Critics of forensic DNA like to point to a Texas case where a murderer was selected out of a small, inbred black population. But in other cases, the logic of "ethnic ceilings" [using ethnic groups as reference populations in order to increase the statistical probability of DNA matches] is wholly irrelevant. In the case of the campground murder, for example, the suspect could have been anyone. When he was identified, it was not because of his race, but by a semen sample. Thus it made no sense to compute the odds *only* against his racial group. Wherever factors other than race have been the key to singling out a suspect, ethnic ceilings on DNA profiles are irrelevant. . . .

Resisting a Tidal Wave

One can't come away from the issue without the impression that the attorneys opposing DNA evidence are trying to hold back a tidal wave of scientific research. Genetics is the most rapidly exploding field in the scientific world. Whatever objections can be raised today will probably be overcome tomorrow. The "polymerase chain reaction" (PCR), a technique that uses a microbe found in hot springs to "amplify" small amounts of DNA, is now being used to make identifications with as little as 20 cells. Experts in the field say the VNTR method may be outdated in the near future. If critics do succeed in having the few private labs taken off the job, their work will be taken over by the FBI and the state crime labs—an outcome that is unlikely to make opponents any happier. At best, defense attorneys can only hope to continue muddying the waters, grasping at every letter-to-the-editor as proof that a "scientific consensus" has yet to be reached.

"DNA-based identification, though highly questionable in its present form, is being sold to a terrified public as a way to solve the heinous crimes we hear about every day."

DNA Fingerprinting Is Unreliable and Inaccurate

Ruth Hubbard and Elijah Wald

Ruth Hubbard is professor of biology emerita at Harvard University in Cambridge, Massachusetts, and the author of numerous books, including *The Politics of Women's Biology*. Elijah Wald is a freelance writer and musician. In the following viewpoint, excerpted from their book *Exploding the Gene Myth*, Hubbard and Wald argue that DNA fingerprinting is a seriously flawed technology that should not be used as evidence in courts. The authors believe that while DNA fingerprinting is touted as an infallible technology, it is actually unreliable.

As you read, consider the following questions:

1. Explain the Castro case and how it exemplifies the authors' reasons for opposing DNA fingerprinting.
2. Why is the assumption that racial groups are homogeneous problematic, in Hubbard and Wald's opinion?
3. Why does the public find DNA fingerprinting an appealing technology, according to the authors?

As reported crime rates soar, so does public discontent with the way the police, investigative agencies, and the courts are handling the apprehension and prosecution of presumed criminals. So, just as the medical system is turning to DNA for a quick fix for people's health problems and the schools are using it to explain children's failure to learn, law enforcement agencies are looking to it for an answer to crime. . . . Law enforcement officials have begun looking at DNA in the hope of developing a system for the positive identification of criminals.

The logic is straightforward enough. Except for the rare individual who has an identical twin, each of us is genetically unique. My DNA is different from that of anyone else and, if it were possible to identify the complete base sequence of my DNA, it could be used to identify me absolutely. Since it is not yet possible to do that, or even to identify individual characteristics of it unequivocally, scientists have come up with approximations—techniques that promise to identify a person to a probability of, say, one in a million or, better yet, a billion.

With such a technique in hand, forensic scientists need only a small sample of tissue—a hair or a spot of dried blood or semen—that a perpetrator has left at the site of a crime, or a victim may have left on the body or clothing of a suspect. They can then compare the DNA-profile of that sample with the profile of a blood sample taken from a suspected perpetrator or from the victim and, ideally, can get a decisive match or an exclusion. In theory, DNA-based profiles are absolute identifiers, like fingerprints, only less subject to deterioration or tampering and more likely to be retrieved as evidence. Advocates of this new procedure call it *DNA-fingerprinting*, though I will avoid that term because at present DNA-based identifications are not nearly as unequivocal as fingerprints can be.

The Risks of Relying on DNA

Like fingerprints, forensic DNA samples can be degraded or contaminated. If, for example, the sample has been collected from clothing or a rug that was recently cleaned with a synthetic detergent, residues from the detergent can change the DNA so that restriction enzymes will cut it differently than they would have cut the original sample. Also, tissue or blood samples can easily be contaminated by bacteria, in which case the bacterial DNA becomes part of the sample and will produce misleading results. Both these problems would lead to false exclusions, while other problems can lead to false matches.

In one criminal prosecution, a man named José Castro was indicted for the murder of a neighbor and her two-year-old daughter in the Bronx, New York. The prosecution claimed that a spot of blood that was found on Castro's watch had been identified by DNA typing as coming from the murdered woman. Lifecodes Corporation of Valhalla, New York, the commercial firm that performed the DNA match, asserted that the DNA pattern of the sample on the watch matched that of the murdered woman and that the odds of finding that pattern in the general population were 1:189,200,000, which made the identification sound pretty decisive.

When the case came into court in early 1989, two lawyers, Peter Neufeld and Barry Scheck, decided to challenge this evidence. They called in Eric Lander, a geneticist and mathematician at the Whitehead Institute in Cambridge, Massachusetts, as an expert witness. Lander was disturbed by the poor scientific quality of the data Lifecodes presented to establish the supposed match. He estimated the odds for a random match, which Lifecodes had said were around one in two hundred million, as one in twenty-four—not a very convincing identification. Then something unprecedented happened: Expert witnesses for both the prosecution and defense met together in New York and after evaluating all the data, they issued a consensus statement in which they challenged the adequacy of the evidence Lifecodes had presented as the basis for its statement that the DNA in the two samples was the same. As a result, the judge disallowed the use of this evidence. (Despite this, and to the dismay of Neufeld and Scheck, Castro pleaded guilty and was sentenced to a lengthy prison term.)

The failure of DNA identification in the Castro case has called into question evidence both Lifecodes and Cellmark Diagnostics, the other major DNA-matching firm in the United States, have given in previous cases. Since the Castro case, DNA-profiles have been disallowed in many state courts, though they have been admitted into evidence in others.

How the Science Works

To understand this controversy, we need to look more closely at how DNA matching is done. Identifiable fragments of DNA, called RFLPs, can be produced by letting restriction enzymes chop up DNA into pieces of different lengths. For purposes of genetic diagnosis, this process can then be used to differentiate between family members who carry a particular allele and those who do not.

The technique used to generate DNA-profiles for use in forensics is similar, except that here scientists are not looking for genes. The method is based on the fact that, for some unknown reason, occasional short sequences of base pairs will be repeated over and over, sometimes more than a hundred times. These repeats lie next to one another (*in tandem*), and our chromosomes all contain stretches of DNA made up of such repeating sequences. Because the chromosomes of different people vary in the number of these repeats, such sequences are called *variable number of tandem repeats* or *VNTRs*. VNTRs appear to be randomly interspersed in the human genome, and are not known to have any biological function. Forensic scientists have settled on analyzing three or four specific VNTRs for purposes of DNA-based identification.

As with other kinds of RFLPs, VNTRs are identified by digesting a sample of DNA with a battery of restriction enzymes. The mix is placed at the top of a gel on which VNTRs of different size travel at different rates. The different VNTRs are then made visible by tagging them with specific radioactive markers. In theory, since different people's VNTRs are of different lengths, they will move at different rates on the gel, so that each person should have his or her own characteristic pattern.

What makes this method particularly attractive for purposes of criminal investigation is that VNTR analysis requires very little material, so that traces of tissue left at the site of a crime are often enough. If they aren't, a technique called the *polymerase chain reaction* or PCR makes it possible to copy minute samples of DNA over and over many times, so that a few molecules of DNA are sufficient to generate a sample large enough for analysis.

The simplicity of the technique and the supposed decisiveness of the identifications it provides have led to attempts to introduce DNA matching in over a hundred recent litigations at the state level, as well as in a few federal cases. It is also being used in Canada, Great Britain, and on the European continent. The FBI has set up its own laboratory for DNA typing at Quantico, Virginia, and private firms such as Lifecodes and Cellmark have set up commercial laboratories. The U.S. Department of Justice is supporting university-based research into VNTR matching and other potential methods of DNA profiling.

Scientific Problems with DNA Profiles

So what went wrong in the Castro case and what are some of the issues that require a more critical look?

There are both technical and theoretical problems with VNTR identifications. For one thing, mistakes can happen and samples can get mixed up, as they apparently did in another Bronx case in 1987. In the Castro case, when Lifecodes compared the gels on which its scientists had run the DNA from the blood sample found on Castro's watch and the DNA isolated from the murdered woman, they decided to ignore the fact that the positions of the VNTR bands did not match up precisely. They rationalized these differences by saying that the gels were just a little different, but they ran no controls to check whether this was in fact the case. It was on this basis that Lander disputed their evidence and the scientific experts on both sides ended up discounting it.

In other words, forensic laboratories sometimes fail to use accepted scientific standards of what constitutes a match. That seems easy enough to repair. But even when used properly, the matching technique is not all that good. When the FBI's Forensic Laboratory ran profiles on samples of DNA prepared from blood drawn from 225 FBI agents, and the tests were repeated a second time with the same samples and by the same laboratory, one in six of the results did not match up.

In another test of the procedures, the California Association of Crime Laboratories sent fifty samples each to Lifecodes, Cellmark, and Cetus Corporation, a biotechnology firm in Emeryville, California, which performs DNA-based identifications by a somewhat different technique than the other two firms. Both Cetus and Cellmark mistakenly matched samples that were not identical. Lifecodes got all the matches right, but it turned out that the test samples had been run by their research scientists rather than by the technicians who normally perform DNA-matches.

There is another technical problem. Let us imagine that a specific VNTR in my DNA consists of 112 repeats, but in my neighbor's DNA the same VNTR has only 109 repeats. That is a genuine difference between us, but the available techniques are not sensitive enough to pick up such small variations and will report our two VNTRs as identical. When technicians compare VNTRs, the larger the number of repeats in a given VNTR, the greater the difference in the number of repeats between two people has to be in order for that difference to be detected.

Fictitious Racial Groups

Even if the matching techniques become more precise, there is a more fundamental and decisive question. Against what reference population is one to judge the probability that a "randomly chosen" individual will not have the same pattern of VNTRs as someone else? What constitutes a genetically "random" population? Two geneticists, Richard Lewontin and Daniel Hartl, have discussed this issue in *Science* magazine.

For identification purposes, the FBI has established reference populations, which they call "Caucasian," "Black," and "Hispanic." Each group is assumed to be homogeneous and people are assumed always to select their mates at random from within their own group. So, by averaging a few samples from each of these populations, reference samples have been constructed that serve as the standard VNTR pattern from which to make statements about the probability of finding a match within that population.

All these assumptions are problematic, but let us just tackle two obvious ones: that the populations are homogeneous and that there is random mating within each. The U.S. "Caucasian" population consists of immigrants from all over Europe, some of whom have arrived recently, some many generations ago. Some have come from small villages where their ancestors lived for hundreds of years, others from large, cosmopolitan cities. In this country, "Caucasians" often live and marry within fairly distinct communities of Italian-Americans, Swedish-Americans, Irish-Americans, and so on. Yet the statement that the odds are less than one in a hundred million that two individuals have the same pattern of VNTRs is based on the assumption of random sampling from a fictitious, homogeneous, and randomly inter-breeding "Caucasian" community.

A Big "If"

DNA fingerprinting may sound like the ideal evidence for an open-and-shut case—if the tests are done accurately. But that . . . is a big "if."

Peter Doskoch, *Science World*, October 21, 1994.

A similar argument can be made about the census category "Black." It may be convenient, but it has no genealogical or bio-logical meaning. U.S.-born African Americans may come from small rural communities in the South, where their ancestors have lived since they were brought to this continent, or from families who moved north and live in industrial centers like Chicago or Detroit. Some "Blacks" have immigrated recently from Barbados or Jamaica, or from one of the African states. How similar genealogically are a "Black" who has recently im-migrated from Trinidad, a second- or third-generation "Black" from Harlem, and a "Black" farmer from Mississippi?

The category "Hispanic" is even less meaningful, since the term includes Caucasians from the United States, Central and South America and the Caribbean, Cuban and Puerto Rican Blacks, and Native Americans from all over Latin America.

The United States has never been, and is not now, a genealogi-cal melting pot. Any model built on the existence of a well-homogenized, randomly mating population is a fiction and bound to fail. An extreme example of this came in *Texas* v. *Hicks*, in which a man was sentenced to death for a murder which he staunchly denied having committed. DNA-matching was said to show he was the murderer, and the testing labora-tory quoted the usual astronomical odds against another person's DNA matching his closely enough to provide a false result. However, as Eric Lander points out, "The crime occurred in a small, inbred town founded by a handful of families," so the probability of finding another person with the same DNA-profile (within the reliability of the technique) must have been considerably greater than was reported.

The Divided Scientific Community

The scientific community itself is divided over the validity of DNA matching. When Lewontin and Hartl submitted their critique of the technique to *Science*, its editor asked Ranajit Chakraborty and Kenneth Kidd, two scientists known to support the technique, to write a rebuttal for the same issue. As is often the case, this is not a purely scientific disagreement. A news article in *Nature* revealed that Chakraborty is a coinvestigator on a $300,000 grant for DNA forensics research from the National Institute of Justice, which is funded by the U.S. Department of Justice. The Department of Justice is by no means a disinterested party in this discussion. It is deeply committed to the use of "DNA fingerprinting," and both *Science* and *Nature* reported attempts by a Department official to derail the publication of Lewontin and Hartl's article. The Justice Department's intervention in this supposedly scientific debate is both inappropriate and frightening. If the value of the technique is still in question, surely it is in the interests of the criminal justice system to resolve the debate rather than to bury it.

Since the publication of the opposing articles, several scientists have written letters to *Science* on both sides, and the original authors have responded in turn. Clearly, disagreement is not dying down. The situation has become so confusing that when a National Academy of Sciences panel in April 1992 issued what was supposed to be the authoritative statement on the value of DNA-based forensic techniques, this report was interpreted in diametrically opposite ways. The *New York Times* ran an article about it under the headline "U.S. Panel Seeking Restriction on Use of DNA in Courts," adding the subheading "Judges Are Asked to Bar Genetic 'Fingerprinting' Until Basis in Science Is Stronger." Then, the very next day, the *Times* and other papers ran an Associated Press dispatch which explicitly contradicted the first article. The AP story quotes Victor McKusick, the chairman of the National Academy panel, as saying, "We think [genetic fingerprinting] is a powerful tool for criminal investigation and for exoneration of innocent individuals and one that should be used even as standards are strengthened." McKusick said that the first *Times* article "seriously misrepresents our findings," but *Times* writer Gina Kolata stands by her story, and my own reading of the report bears her out.

135

As usual, there was also a question of conflict of interest. Several members of the National Academy panel serve on boards of directors or have financial interests both in genetic screening companies and in companies involved in "DNA-fingerprinting." One such member, C. Thomas Caskey, resigned from this panel in 1991 after a *Nature* article revealed his financial links to a company doing DNA-based identifications.

The Power of Science

What all this boils down to is that, at present, the reliability of the data and of the scientists producing those data are in doubt. Unfortunately, juries and judges, like the rest of the public, are easily swayed by the mystique and power of science. When it is offered with appropriate fanfare, a DNA match need not be scientifically reliable to prove decisive in a court of law. . . .

In the past, the FBI has consistently obstructed the introduction of regulations and standards for DNA matching. It has opposed independent testing of its own results, as well as proposals to require laboratories to document their conclusions in writing and to have the scientists and technicians who perform the tests sign their reports. As a result, at present, "no private or public crime laboratory [in this country] . . . is regulated by any government agency," so that "there is more regulation of clinical laboratories that determine whether one has mononucleosis than there is of forensic laboratories able to produce DNA test results that can help send a person to the electric chair," according to Peter Neufeld and Neville Colman.

Despite the problems and disagreements, law enforcement officials in many areas are going ahead and using DNA technologies, and are beginning to store tissue samples and set up data banks of "DNA fingerprints." In an article published in *Parade Magazine* in March 1991, Earl Ubell writes that "several states now are taking blood samples from convicted rapists and other violent criminals. Their DNA profiles will be stored in a data bank for use by police across the United States." In support of this practice, he writes: "Using DNA fingerprinting, for example, detectives could trace a rapist convicted in Utah who later rapes in Ohio by matching the DNA 'prints' on file with those in traces found on the victims."

So, DNA-based identification, though highly questionable in its present form, is being sold to a terrified public as a way to solve the heinous crimes we hear about every day. This is a quick fix, rather than a real solution. Blaming yet another crime on a convicted rapist or murderer may make law enforcement officers look (and feel) better but, if the charge is built on unreliable evidence, it does not make us any safer on the street or in our homes.

"*DNA testing has great implications for exonerating criminal suspects as well as convicting them.*"

DNA Fingerprinting Is Gaining Acceptance

James H. Andrews

DNA fingerprinting is a powerful tool for convicting criminals and protecting innocent suspects, James H. Andrews writes in the following viewpoint. While defense attorneys have attacked the accuracy of labs that perform DNA testing, Andrews reports, improvements in the scientific techniques will lead to more widespread acceptance and use of the technology in the near future. Andrews is a staff writer for the *Christian Science Monitor*, a daily newspaper.

As you read, consider the following questions:

1. In the author's view, how has the development of PCR technology improved DNA fingerprinting?
2. What are the limitations of DNA evidence as compared to actual fingerprints, according to Andrews?
3. What do most critics of DNA fingerprinting think about the science behind the technology, according to Andrews?

James H. Andrews, "Genetic Fingerprinting Revolutionizes Police Work," *The Christian Science Monitor*, June 20, 1994. ©1994 by the CSPS, Inc. Reprinted with permission.

When Nidal Ayyad sealed the envelope, the act helped seal his fate.

A forensic analysis of saliva on an anonymous letter to the *New York Times* after the World Trade Center bombing in 1993 linked Mr. Ayyad to the terrorist attack. The saliva test was only a small part of the evidence that eventually led to the conviction of Ayyad and three other men for the New York City bombing, but it illustrates the vast crime-solving potential of genetic biology.

From the envelope flap, Federal Bureau of Investigation laboratory examiners extracted a sample of Ayyad's DNA—deoxyribonucleic acid, an organic substance in cells. Each person (except identical twins) has a unique DNA makeup. Matching the DNA sample from the envelope with a sample obtained from Ayyad, law-enforcement officials identified him as the writer who claimed credit for the bombing.

Because blood and semen are the DNA-bearing cellular materials most commonly found at crime scenes or on crime victims, DNA testing is often used in cases of murder, rape, and other violent crimes. But it has wider forensic applications as well. "Even a burglar who scratches himself breaking a window leaves behind a DNA calling card," says John Hicks, director of the FBI's laboratory division in Washington.

A Profound Development

Thanks to a recent advance in DNA technology called polymerase chain reaction (PCR), which lets researchers reproduce the DNA in minute samples found at crime scenes, scientists "can get very probative information from DNA samples that formerly were just too small to examine, like saliva from cigarette butts or the back of postage stamps, and even the DNA in the root of a single human hair," says Paul Ferrara, director of the state of Virginia's Division of Forensic Science.

Dr. Ferrara calls DNA analysis "a profound development in forensic science, even greater for identifying individuals than the use of fingerprints starting in the last century.

DNA testing has great implications for exonerating criminal suspects as well as convicting them. According to Ferrara, about 25 percent of the tests his department performs absolve suspects. Barry Scheck, a professor at Cardozo Law School in New York and a criminal-defense lawyer, heads up an "Innocence Project" at the law school. "We have used DNA evidence to get a lot of people out of jail, including a prisoner on death row," Mr. Scheck says. . . .

DNA testing procedures are too complex to explain in a short article. In summary, after several steps to extract DNA from cells, slice it into fragments of various lengths, and arrange the fragments in patterns, examiners can display the patterns on X-ray film in lanes that resemble crude supermarket bar codes. They can then compare patterns of DNA from a crime scene or victim with DNA obtained from a suspect or located in a DNA databank, in search of a match, or "hit."

Possible Errors

This analysis of cell fragments to single out criminal suspects is sometimes called "DNA fingerprinting." Forensic experts prefer the terms DNA profiling or identification, however. For fingering suspects, DNA analysis has the potential to be more effective than actual fingerprints; but also, as currently obtained, DNA evidence has certain limitations less applicable to fingerprints:

• Although performing forensic DNA tests isn't akin to rocket science, it requires technical precision by skillful examiners. Mistakes occur, both in the chemical procedures and through such human errors as mislabeling vials containing the DNA samples.

• While each person's complete DNA profile is unique, the small segments of a person's DNA chain capable of being forensically examined are not necessarily unique. Certain DNA patterns recur in humans' genetic makeup, with degrees of frequency determined partly by such factors as race, ethnicity, or inbreeding among groups.

Consequently, once they have matched a suspect's DNA with a DNA sample from a crime scene, forensic examiners, drawing on data from wide population samples, must calculate the probable frequency with which the identified DNA pattern would crop up in different humans. Thus, an examiner at a trial might testify that a suspect's DNA profile would be repeated in 1 in a million people, or 1 in 100,000, or 1 in just several hundred.

A debate has raged in the scientific and criminal-justice communities over the representativeness of the population samples used to make such estimates and the proper methods of calculation.

Defense lawyers have seized on both factors—possible testing errors and the dispute over genetic population studies—to challenge DNA evidence in court.

According to Ferrara, the debate over population statistics— that is, the significance of getting a DNA match—is dying down. The toughest challenges from defense lawyers, in his experience, are on the accuracy of specific DNA tests and, more broadly, on the adequacy of DNA quality-assurance standards and safeguards in forensic labs.

Even Critics Praise Use of DNA

Professor Scheck and New York City attorney Peter Neufeld, the co-chairmen of the National Association of Criminal Defense Lawyers' DNA task force, assert that quality-assurance standards are lax in many forensic labs that perform genetic testing.

Getting a Match

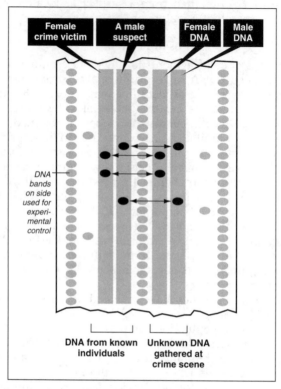

Forensic scientists attempt to match DNA evidence gathered at a crime scene with known DNA collected from identified individuals (victims, suspects).

Through a process called restriction-fragment-length-polymorphism (RFLP) analysis, scientists break DNA samples into fragments of varying lengths and arrange them into banded patterns in lanes on radiation-sensitive film, called an "autoradiograph."

In the depiction above of an autoradiograph from a murder case, DNA samples from known sources are compared with samples from unknown or questioned sources found at the crime scene, such as blood stains, semen, or hair. The bands in the victim's sample match the bands in the unknown female's sample; similarly, the suspect's sample matches the male portion from the DNA crime-scene evidence. This genetic match means there is a high—but not perfect—probability that this suspect was at the scene of the murder.

Source: FBI Laboratory Report, 1993.

They call for more "blind" (disguised) proficiency testing of labs' DNA procedures by outside examiners, and they argue that labs' error rates should be calculated and presented to juries. . . .

Not even critics like Scheck and Mr. Neufeld disparage the underlying DNA science (indeed, they praise its application to freeing innocent people). In 1991, a committee of genetic scientists and forensic experts established by the National Research Council, National Academy of Sciences, reported after a two-year study that DNA testing "is a fundamentally reliable technology when properly applied by qualified people," in the words of Ferrara, a member of the panel.

Clearly, DNA profiling is here to stay. In Ferrara's large and well-funded forensic department, DNA testing has totally replaced traditional serological tests (such as comparing blood types), and in short order most other labs will catch up, especially if federal grants are forthcoming, he says.

By 2000, he predicts, law-enforcement officials will use and exchange DNA evidence as routinely as they utilize fingerprints today.

"DNA fingerprinting, hailed as a powerful force of justice, can never be the final authority."

DNA Fingerprinting Is Problematic

Pat Spallone

Pat Spallone is a freelance researcher and writer associated with the Centre for Women's Studies at the University of York in Great Britain. A former biochemist, she is the author of the books *Beyond Conception: The New Politics of Reproduction* and *Generation Games: Genetic Engineering and the Future for Our Lives*, from which the following viewpoint is excerpted. In the viewpoint, Spallone cites several cases in which the DNA evidence used in court cases was shown to be flawed. The author believes that inaccurate DNA results threaten the freedom of innocent defendants.

As you read, consider the following questions:

1. Why is it difficult to get reliable results from DNA fingerprinting, in Spallone's opinion?
2. How might women be harmed by the use of DNA fingerprinting for paternity testing, according to Spallone?

DNA fingerprinting is used mostly in police work, or to prove familial relationship. . . . At first, DNA fingerprinting seemed failproof and advisable, but the circumstances of its use proved more complicated and troubling than was anticipated. For me these raised three issues: about confidence in the ability of science to create certainty from mystery; about how we live in relation to technology; and about authoritarianism and human identification technology.

State of the Art

DNA fingerprinting or genetic fingerprinting was a chance discovery of English geneticist Alec Jeffreys. In the early 1980s he was working at the private Lister Institute, studying regions along the DNA known as 'hypervariables'. He recognised that each person has a unique pattern to these regions, and that he could create 'gene probes' to identify them.

An individual's DNA is easily available for analysis, since most cells in the body carry a complete set of chromosomes, where the DNA is packaged. Any blood or tissue sample will do. The DNA is extracted; enzymes are used to chop the DNA into fragments. The fragments are separated out from each other according to (molecular) size. Gene probes are added which identify the hypervariable regions; the resulting profile appears as a series of dark bands on a gel or X-ray film surface.

Despite the connotation of the term 'DNA fingerprint', the profile obtained by Jeffreys' method is not of the total genetic material in a person's cells, and it is not a profile of 'genes' (pieces of DNA that code for proteins). The 'hypervariable' regions of DNA were discovered in 1980, and scientists do not know what their function is. Jeffreys' technique is only one type of gene profile. In principle, any part or the totality of a person's DNA can become a genetic profile. But that is a principle, which is why Jeffreys' discovery was so scientifically and commercially newsworthy.

The DNA fingerprint is said to be specific. The distribution of 'hypervariables' is unique for each person: they are inherited: half are shared with the mother, half with the father. The chances that two people will possess the same DNA fingerprint pattern is thirty billion to one according to Cellmark Diagnostics, a company formed to market the method. Jeffreys calculated that the chances are three times greater. Hence, the 'fingerprint' can identify an individual, or a genetic family relationship one or two generations apart (that is among parents, grandparents and children). Jeffreys stressed that there was no other sort of information hidden in the pattern, like race or hair colour.

The DNA test was patented by Jeffreys and the Lister Institute; within a few months the UK chemical giant ICI bought the world rights and created Cellmark Diagnostics with locations in Abingdon, England, and in the state of Maryland, USA. ICI expects to earn millions from the method. Variations on the Jeffreys test from other researchers and new companies followed in quick succession. One commentator suggested that in future, DNA tests would be designed to tell the colour of eyes, race, and other identifying physical features. . . .

No Uniform Methods

DNA identity tests are increasingly coming under attack. In Massachusetts, the state's Supreme Judicial Court ordered a new trial for a man convicted of raping a 14-year-old disabled girl, saying that the scientific community does not have widely accepted uniform methods for testing and interpreting DNA evidence.

Simson L. Garfinkel, *The Christian Science Monitor*, March 27, 1991.

In England and Wales, DNA fingerprinting was first used in a forensic capacity to identify an individual whose blood or semen samples were left at the scene of a crime. In January 1987 police in Leicester, England, sought the killer of two teenage girls through mass screening of the men of two villages. They asked the 2000 local male residents to come forward voluntarily for elimination purposes. The murderer, Colin Pitchfork, was found in January 1988, but only after some confusion, since he got another man to give a blood sample in his place. He was given a life sentence.

The first time DNA fingerprinting was entered as evidence in a jury trial, on 5 October 1987 at Birmingham Crown Court in a rape case, problems with the technique began to emerge. A woman identified two men as the rapists. One was convicted and imprisoned for five years, and the second man was acquitted. The DNA fingerprint came from semen left on the woman, but it proved confusing to the case. The DNA fingerprinting method was not sensitive enough, said Home Office forensic scientist Dr David Werrett. He added that a forensic sample is not the perfect working sample of clinicians who can get fresh, large samples of blood or tissue; at this point, he added, the evidence from conventional fingerprint tests were just as conclusive as genetic 'fingerprints'. He said the tests needed to be improved.

The use of DNA fingerprinting in this rape case demonstrated that the DNA association of a man's semen is not going to change the terms of rape trials. What it does provide is another type of identifying evidence. If the test is considered conclusive in identifying a man accused of rape, it provides evidence for the conviction of an accused rapist, as it did in a rape trial in Wales the following month. . . .

Questionable Results

If it seems that DNA fingerprints are necessary for responsible policing because they are super accurate and too useful to give up, consider the following incidents which undermined the scientific certainty of genetic identity technologies.

In the USA, a number of cases of questionable results arose as the use of DNA fingerprinting in police work increased. The first was a murder investigation in the Bronx area of New York City. A woman and her daughter had been killed. A local man, Jose Castro, was interrogated by the police acting on a tip. Lifecodes, a company carrying out the DNA tests, concluded that a bloodstain found on the watch of Jose Castro matched the DNA patterns of the woman. Lifecodes experts calculated that the frequency of the pattern in the Hispanic population was about one in 100,000. But fundamental problems with the assessment were uncovered—very technical details concerning the analysis of the blood sample—and what had seemed conclusive evidence could no longer be considered so. Eric Lander, the geneticist who analysed the data for the counsel for the defence, added that 'The general issues of interpretation are unique neither to the Castro case nor to Lifecodes.'

Lander described three additional cases in which DNA fingerprinting would have brought about miscarriages of justice. One was of a woman accused of abandoning her newborn baby who was found dead, allegedly in the back seat of her car. Cellmark Diagnostics reported that its DNA analysis showed that she was the mother. She was not, as eventually became clear. A medical examiner determined that the baby was stillborn on 4 February 1988. The woman accused could never have given birth to that baby because she was in the early stages of pregnancy at the time, and gave birth the following October.

Lander pointed out several problems with DNA tests used in forensics. Someone must determine if two unknown samples are identical. Reliability in recognising a match is technically demanding. The DNA fingerprinting results being used in criminal court cases in the USA are based on flimsy, inconclusive judgments. DNA fingerprinting also depends on inferences about the frequency with which matching patterns will be found by chance, which in turn rest on simplifying assumptions about population genetics. Contaminated probes are often used. 'It is my contention that DNA forensics sorely lacks adequate guidelines for the interpretation of results,' Eric Lander concluded.

DNA fingerprinting, hailed as a powerful force of justice, can never be the final authority. These cases shook up those who had previously been so convinced of the accuracy and reliability of DNA fingerprinting. They also suggest that the new technology is not always necessarily going to serve women, although one of the other main targets of the new technology was paternity testing, hailed as a force of justice for women.

Policing Paternity

Cellmark saw paternity testing as its most lucrative application and let it be known in women's magazines and the media that 'Women left holding the baby now have technology on their side.' In the six months between June and December 1987, the Abingdon laboratory dealt with about 1800 cases to identify familial relationship, at £120.75 [about $215 U.S.] a sample. Thirty per cent of these were paternity cases, and 60 per cent immigration cases.

What bothered me during all the media coverage of these two uses of gene profiling was that no one questioned the genetic concept of paternity and maternity that underlay the use of DNA fingerprinting to prove the parent-child relationship. This particular biological definition (a genetic one) is a confined definition of the family, of parenting and child caring. For one, what of children who may have been taken into a home, or adopted? Are they not real children? Or are they 'less' real because they are not biologically related by shared DNA?

While DNA fingerprinting might help identify some errant genetic fathers and make them pay child support, it is fraught with problems. If an individual woman wishes to make claims on the father through DNA tests, it looks as if she will have to foot the bill. It could also work against women. What repercussions would there be on women if the father wants to use DNA tests to see if in fact he is the genetic father with legal obligations towards individual children; or to claim custodial rights to children against the wishes of the mother? What if DNA testing revealed that the putative father is not the genetic father? On a policy level, it may put pressure on the man to assume a social role in the life of the woman which she does not want. Governments may embrace it to force errant fathers to take financial responsibility rather than for the state to provide support for children. This solution may not be a solution for the women and children; and on the level of social policy, it is regressive in that it expects women to be economically dependent on men by definition rather than enabling women to live their lives autonomously. The point is, DNA fingerprinting is presented as doing favours for women as a whole when in fact it rests on patriarchal assumptions which have in part caused the problems it is supposedly solving.

Periodical Bibliography

The following articles have been selected to supplement the diverse views presented in this chapter.

Shannon Brownlee	"Courtroom Genetics," *U.S. News & World Report*, January 27, 1992.
Shannon Brownlee	"Science Takes a Stand," *U.S. News & World Report*, July 11, 1994.
Howard Cooke	"DNA and Police Files," *The Lancet*, July 17, 1993. Available from 42 Bedford Sq., London WC1B 3SL, England.
Peter Doskoch	"DNA on Trial," *Science World*, October 21, 1994. Available from 555 Broadway, New York, NY 10012-3999.
David E. Houseman	"DNA on Trial—the Molecular Basis of DNA Fingerprinting," *The New England Journal of Medicine*, February 23, 1995. Available from 1440 Main St., Waltham, MA 02254.
Leon Jaroff	"Order in the Lab!" *Time*, August 8, 1994.
Kenneth Jost	"Science in the Courtroom: Is Scientific Evidence Being Misused in Lawsuits?" *CQ Researcher*, October 22, 1993. Available from 1414 22nd St. NW, Washington, DC 20037.
Kevin Krajick	"Genetics in the Courtroom: Controversial DNA Testing Can Clear a Suspect," *Newsweek*, January 11, 1993.
Mark Nichols	"DNA on Trial," *Maclean's*, February 6, 1995.
Paul M. Rowe	"DNA Evidence on Trial in USA," *The Lancet*, December 3, 1994.
Jim Schefter	"DNA Fingerprints on Trial," *Popular Science*, November 1994.
Pamela Zurer	"DNA Profiling Fast Becoming Accepted Tool for Identification," *Chemical & Engineering News*, October 10, 1994. Available from 1551 16th St. NW, Washington, DC 20036.

How Will Genetic Engineering Affect Health Care?

GENETIC ENGINEERING

Chapter Preface

In recent years, scientists have discovered what they believe to be the genetic causes of a variety of diseases and conditions, from cystic fibrosis to alcoholism. Identifying which gene is related to a specific disease is a monumental discovery, for once physicians know this, they can create tests to identify individuals who carry the gene in question and can devise gene therapies to treat or perhaps cure the disease.

Thomas F. Lee, the author of *Gene Future: The Promise and Perils of the New Biology*, describes the two types of gene therapy:

1. *Somatic-cell gene therapy* aims at introducing genes into some of the somatic (body) cells to correct a genetic defect. . . .

2. *Germ-line gene therapy* would result in the correction of the genetic problem in an individual's reproductive cells so that it would not be passed on to his or her offspring.

While germ-line gene therapy is not yet a reality, according to Lee, researchers are currently devising or attempting to develop somatic-cell gene therapies for a variety of genetic disorders, including cancer, cystic fibrosis, muscular dystrophy, and AIDS.

But while genetic engineering may hold immense potential for treating disease, many critics believe that this new technology comes at a cost. Most of these critics primarily fear the resurgence of a eugenics movement, in which the government would force people to change their genetic makeup or that of their children in order to create a class of "perfect" humans. For example, the government might test all citizens for a particular "defective" gene, then prohibit those with the gene from having children. As an editorial in the *Christian Science Monitor* states, "As knowledge about human genetics expands, so does the temptation to label as abnormal or as maladies traits that today are seen as 'normal' physical variations. . . . The idea that genetics can explain and address many of humanity's physical and behavioral ills threatens to reinforce a notion that humans are merely the sum of exchangeable molecules. Such reductionism invites intolerance and social engineering."

The thought of a new eugenics movement is frightening to many. But is it right, or even possible, to reject the enormous benefits that might come from gene therapy? The following chapter examines these critical issues.

"Biologists today believe that by tinkering with people's genes . . . they will eventually be able to eliminate most of the diseases that now plague the world."

Genetic Research Will Improve the Quality of Health Care

The Economist

In the following viewpoint, the editors of the *Economist* express the belief that the knowledge gained from genetic research will benefit humankind and improve the quality of health care. According to the editors, experts predict that most common serious genetic diseases will be eradicated within the next fifty years. The *Economist* is a weekly British publication that covers business, financial, and political issues.

As you read, consider the following questions:

1. Explain the medical advance made by physicians French Anderson, Michael Blaese, and Ken Culver in 1990, as described by the authors.
2. What advances in gene therapy have been made by the companies Genentech, Amgen, and Oncor, according to the *Economist*?
3. How could gene doctors "shape human destiny," in the authors' opinion?

The Economist, "Engineering Health," March 19, 1994, ©1994, The Economist, Ltd. Distributed by New York Times Special Features/Syndication Sales. Reprinted with permission.

Computers, telecommunications and robots may make doctors and hospitals more efficient and safer. Biology will take medicine to places that are not even dreamed of—yet. Since the early 1970s scientific discoveries have turned biology from being a discipline dedicated to the passive study of life into one that can alter it at will. Biologists today believe that by tinkering with people's genes, the units of heredity, they will eventually be able to eliminate most of the diseases that now plague the world. Tomorrow, such extraordinary ambitions may seem modest, as scientists start to work on improving a person's genetic lot in life.

The Work of Genentech

It all started in the early 1970s, when scientists first learnt how to clone and engineer genes. In cloning, a single gene is isolated from millions of others. Before this, scientists were confronted with the genetic equivalent of noise. Now they were free to study the structure and function of gene entities in isolation. By the end of the decade, Genentech, in San Francisco, had launched the first-ever genetically-engineered drug, human insulin. What Genentech had done was to take the cloned gene coding for human insulin and transfer it to bacteria. Genentech had synthesised a new life-form, a bug capable of making a protein foreign to itself. For centuries selective breeding has produced novel crops or cattle, but always with unpredictable results. With genetic engineering, scientists can be surer of outcomes: that a particular bacterium will produce insulin, say.

Scientists now have a rag-bag of new tricks to help them probe nature. Mike McCune of SyStemix in Palo Alto, California, an experienced geneticist, points to four other bits of cleverness crucial to the progress of biotechnology, as the new field of biology became known. On the McCune list are the cloning of pure antibodies, polymerase chain reaction, differential hybridisation and multiparameter flow cytometry. Without going into the details of what this jargon means, all four aim broadly at the same goal: to provide a better understanding of what makes nature tick. This knowledge is now being put to good effect, with the discovery of powerful new medicines.

Biotechnology has made big promises before, without delivering on its early hype. But as Glaxo's Sir Richard Sykes points out: 1993–1994 has seen a paradigm shift in modern biology, because it is revealing so much information about the basic mechanisms of disease for which drugs can be developed. In the past, pharmaceutical firms relied on serendipity to find new drugs. In the future that is not the way to go if the idea is to produce medicines of value.

The rest of the drug industry feels the same way. According to

Steve Burrill, a biotechnology buff, in 1993 drug firms formed around 100 strategic alliances with small biotech firms to tap into their know-how—twice as many as in the previous year. Research successes have fuelled a huge expansion of the biotechnology industry. In 1993 Mr Burrill counted 1,300 biotech firms in America, 200 in Britain, and 400 elsewhere in Europe. Mr Burrill reckons that by 2010 biotechnology firms' sales will have grown ten-fold compared with 1993, to some $100 billion. Because of the long lead-times involved, products began to trickle out of biotech R&D laboratories only about 1989. Two drugs already have sales in excess of $1 billion a year, because they are so good at what they do: Amgen's EPO, which prevents anaemia during kidney dialysis, and its Neupogen, which decreases the incidence of infection in cancer patients undergoing chemotherapy. But these two seem dull compared with what the next generation of biotech products will bring.

Views differ about where biotechnology's biggest contribution will be made. But for a 2010-plus outlook the overwhelming vote goes to human genetic engineering.

The Gene Genie

On September 14th 1990, after years of foot-dragging, America became the first country to allow new genes to be introduced into people. On that day French Anderson, Michael Blaese and Ken Culver, all at the National Institutes of Health (NIH), used a gene drug to treat a four-year-old girl with severe combined immunodeficiency (SCID), a rare and dreadful disease, whose sufferers once had to live inside a sanitised plastic bubble. Those with SCID lack a gene that controls the production of an enzyme known as adenosine deaminase (ADA), which plays an important role in the body's immune defences. Dr Anderson put copies of the ADA gene into the girl's white blood cells. In early 1991 a nine-year-old girl with ADA deficiency was also treated under the gene therapy programme. In May 1993 the two young girls appeared at a press conference looking happy and healthy. The striking results achieved in these two cases have spurred on the use of "gene drugs".

ADA deficiency is one of 4,000 known disorders that result from a single genetic flaw. Most are as rare as SCID; a few, such as cystic fibrosis, are quite common. "But the grand strategy of gene therapy", says an NIH booklet, "also envisages a much broader use of the new techniques to include assaults on heart disease, diabetes and other major health problems that are influenced by the functioning genes." The development of such diseases depends on how a person reacts to environmental factors, such as pollution or smoking. However, the body's susceptibility to them is imprinted in a mix of bad genes inherited from parents. Gene therapy tries to correct these genetic faults to abolish or at least reduce the spread of disease. Dr Anderson, now at the University of Southern California School of Medicine in Los Angeles, says that "it can be used to treat disease, but its primary value will be in prevention." Genetic screening at birth can tell what diseases a person is susceptible to—so genetic protection can be given to prevent the diseases appearing in later life.

Evolution in Action

Genetic differences exist and we're going to have to learn to live with them and to try to improve the lives of those who've been treated very badly by mistakes in DNA replication. But it will always be scary to think that you're different in a way that puts you at a disadvantage. So many people want to deny the fact that such differences exist. Most extreme was the situation in Stalin's Russia, where conventional genetics was banished under the edict that all human differences were due to the environment as opposed to genes.

On the other hand, you must realize the positive effects. If you can tell a woman with a family history of breast cancer that she's not at risk, that is an enormous benefit. It is also a benefit to a woman who learns that she might be at risk and then does something about it. So there's an inevitability to further accumulation of genetic knowledge. We have to get it. But it's going to be very complicated dealing with it. I suspect we underestimate the effect genetics is going to have on medicine and on human life. What we're witnessing is evolution in action. And evolution often isn't kind to the individual. That's why it's so important to increase our programs on ethics.

James D. Watson, *Issues in Science and Technology*, Fall 1993.

In 1993 a lot of progress towards this goal was made. The Centre d'Etude du Polymorphisme Humain (CEPH), in Paris, published the first genetic map of a human genome, the totality of human DNA. Before then, only 2% of the genome had been mapped. What the French did was to establish landmarks (marker genes) among the 100,000 genes that stretch along the human genome. This helps to track down genes that cause most inheritable diseases; patients suffering from such diseases often also inherit distinctive marker genes that are absent in healthy people. With the new map researchers can quickly isolate genes closely associated with the markers to determine which ones cause a disease. The map will also help to obtain a more detailed account of the human genome itself. Thanks to the global efforts of the

Human Genome Project, it is hoped that by 2010 the structure and function of almost all human genes will be understood. Even without the map, in 1993 the genetic causes of several diseases were found. A gene that leads to Huntington's disease, a form of dementia, was found after years of searching. Scientists are close to tracking down genes that cause breast cancer. Tests to screen several diseases were also invented in 1993. Oncor, a tiny biotechnology outfit in Gaithersburg, Maryland, launched a genetic testing service for breast cancer that uses computers to interview patients about family cancer history. It also screens for gene markers associated with breast cancer. As soon as genes that cause breast cancer are found, these too will be screened for to predict a person's chances of contracting the disease. Those with bad test results could opt to take the radical step of having their breasts removed. In time, however, they may get gene drugs that prevent the disease from occurring altogether.

Improving Delivery Systems

Some 250 patients are now being treated with 12 different gene drugs in 74 approved trials around the world: the majority are for cancer, the rest for single-fault genetic diseases, which include haemophilia. The results of several trials are trickling in. Patients with abnormally high levels of cholesterol have, after receiving gene drugs, seen their cholesterol levels fall. Three out of eight patients with terminal brain cancers have experienced a reduction in the size of their tumours. Nobody has reacted adversely to any of the drugs, except for one cystic-fibrosis patient who had breathing difficulties for a few hours. That may have been because the drug was administered through the trachea.

Such technical problems of delivery, which is currently laborious and painful, are slowly being sorted out. There is still some worry about the safety of the delivery system, because viruses, even though inactive, are involved. The virus is a vehicle that carries gene drugs to cells in patients' bodies. Apart from the treatment of cystic fibrosis, the therapy has been administered by extracting bone-marrow cells from the patient, treating them with a virally-packaged form of the gene drug in the laboratory, and then returning them to the body. In 1993 Vical, a biotechnology firm in San Diego, California, found that by combining fat with DNA, it could bypass this procedure and inject genes direct into the bloodstream, much like any conventional drug. Dr Anderson is also working on injectable gene drugs. Researchers are trying to refine delivery systems so that they are longer-lasting and require only a single shot in a lifetime. It is still early days to be sure of the results, but most of the signs emerging from the research are encouraging.

So encouraging, indeed, that Daniel Cohen of CEPH reckons

that by 2010 gene doctors will have found a way of dealing with most diseases caused by single gene defects. Over the next 50 years most common serious diseases will also succumb to gene therapy. And 50 years, he adds, is almost no time at all in the history of medicine (penicillin is now 50 years old). These are inspiring goals.

More controversially, gene doctors also want to shape human destiny. So far they have confined themselves to delivering genes to the somatic cells that make up most of the body. Germ cells in the testes and ovaries are not affected; the new genes are not passed on to the next generation, which remains as vulnerable as its parents to disease. But germ-line gene therapy would correct a genetic defect in the reproductive cells of a patient; offspring would also be corrected and disease could be eradicated.

Improving People

Human genetic engineering could also enhance or improve "good" traits—for instance an extra copy of the human-growth-hormone gene could be added to increase height. On December 31st 1993 a scientific journal, *Nature Genetics*, published an article by CEPH that examined two genes in a group of 338 French people over 100 years old. They found that the centenarians had different levels of genes compared with younger people. A person carrying the right gene variants had twice the chance of reaching old age. This was the first time that genes had been linked to longevity. It follows that gene therapy might extend life-expectancy. And though scientists still do not understand enough about the genetic processes that make humans intelligent or beautiful, it might eventually be possible to tailor people to taste.

"It is misleading for proponents of the genome project to promise that knowing . . . all the genes . . . will lead to cures for a wide range of diseases."

Genetic Research May Not Improve the Quality of Health Care

Ruth Hubbard and Elijah Wald

Genes are related to many serious illnesses, but social and environmental factors are also crucial in understanding and treating such illnesses, Ruth Hubbard and Elijah Wald maintain in the following viewpoint. Hubbard and Wald argue that genetic research will not deliver as many medical miracles as promised. Hubbard is professor of biology emerita at Harvard University in Cambridge, Massachusetts, and the author of numerous books, including *The Politics of Women's Biology*. Wald is a freelance writer and musician. This viewpoint is excerpted from their book *Exploding the Gene Myth*.

As you read, consider the following questions:

1. How do the authors refute the idea that the knowledge gained from the Human Genome Project will bring about cures for severe genetic diseases?
2. Why might an overemphasis on the role of genes make people less healthy, in the opinion of Hubbard and Wald?
3. Why does the ability to "improve" human genetics pose a threat to a society that has a market-driven economy, according to the authors?

Our culture tends to regard health and illness as biological phenomena, but our health is not simply a matter of biology. Social and economic circumstances affect our body states and also shape the ways we perceive and categorize them. Biology cannot be separated from social and economic realities, because they are intertwined in complex ways and build upon each other. We cannot isolate the biological factors, and when we try we oversimplify and distort reality. . . .

Individualization of Health and Illness

While medical scientists often reject social explanations of biological states, they can be more accepting of biological studies that purport to explain social conditions. For instance, Daniel Koshland, a molecular biologist and the editor of *Science* magazine, prophesies that the Human Genome Project will "aid the poor, the infirm, and the underprivileged," because it will improve physicians' abilities to diagnose, and presumably cure, "mental illness." In a *Science* editorial, Koshland states, as though it were an uncontested fact, that mental illness is "at the root of many current social problems" and that understanding the human genome will enable us to move beyond "the current warehousing or neglect of these people."

In this analysis, Koshland not only is making false promises but is actively drawing attention away from the economic and political realities that victimize people. He is diagnosing the poor and underprivileged as sick, and touting the better understanding of genes as a cure for what are clearly economic, political, and social ills.

Koshland is giving a new twist to a very old idea: that people are poor, or rich, because of something inside them rather than because of social inequities. In the nineteenth century, when the science of genetics was born, it was commonly said that "blood will tell." Fictional heroes like Oliver Twist exhibited the virtues and honesty of their middle-class parents, despite their cruel, working-class upbringing. The merchant class attributed its success to a natural superiority over those who failed to rise from poverty. Now, geneticists have translated such perceptions into scientific terms. . . .

Genes as Blueprints

Inherited factors can have an impact on our health, but their effects are embedded in a network of biological and ecological relationships. Genes are part of the metabolic apparatus of organisms that have multiple, mutual relationships with their environments. We breathe our "environment," eat it, sweat and excrete into it, move through it and with it.

This is one reason that even "simple" Mendelian conditions

[regulating inheritance] exhibit varying degrees of severity. Concepts like "the organism," "the gene," or "the environment" are useful as ways to organize our understanding of the world, but we must keep in mind that they do not describe the world as it is. They merely serve to separate out the specific aspects on which we want to focus our attention.

Genes affect our development because they specify the composition of proteins, but it is more realistic to think of genes as participating in various reactions than as controlling them. Because of their complexity and their ability to adapt to change, organisms can sometimes develop ways to compensate for the failure of specific reactions to take place, or for reactions that occur too rapidly or slowly. So, when molecular biologists speak of genes as "control centers" or "blueprints," this is testimony to the hierarchical models they use rather than a description of the ways in which organisms function.

Different Effects

Each protein, and therefore each gene, can affect many of an organism's traits. Conversely, each trait receives contributions from many proteins, hence from many genes. For example, when the gene that specifies the structure of human growth hormone (a protein) was transferred into a mouse embryo, the mouse grew to twice its normal size. When the same gene was inserted into the embryo of a hog, that animal's size did not change, but it became leaner than normal.

The ways this gene functioned depended on other things going on in the organism. To say that the gene "caused" the effects dodges the question of why these effects were different. Obviously, the gene played a part in both cases. Equally obviously, it was not the only factor. Molecular biologists emphasize the role of genes in this situation because they are more interested in genes than in the development of mice or hogs.

Within a single species, as well, the same gene can contribute to different effects in different individuals. In very few cases can a gene legitimately be said to be "for" any one thing. Scientists now know the precise molecular structure of the allele associated with sickle-cell anemia, and for several decades they have known the specific molecular change in sickle-cell hemoglobin that is responsible for this condition. Yet, this knowledge has not enabled them to understand why some people who have sickle-cell anemia are seriously ill from earliest childhood, while others show only mild symptoms later in life, nor has it helped produce cures or even effective treatments. The best medical therapies for people with sickle-cell anemia still rely on antibiotics that control the frequent infections that accompany the condition.

It is misleading for proponents of the genome project to promise that knowing the sequence and composition of all the genes on the human chromosomes will lead to cures for a wide range of diseases. It is all too easy to find proteins associated with specific health conditions, and with present techniques it has become possible to identify genes that specify the composition of these proteins. Such discoveries can be useful, in that they may make it possible to produce large quantities of the proteins, which will make it easier to do research on these conditions. However, this will not necessarily identify their "causes" or cure them.

Only rarely can information at the level of DNA sequences be readily translated into useful information at the level of cells, tissues, or whole organisms. In the past, scientists have deduced the presence of genes, as well as their functions, by looking at the ways organisms differ from one another. There is no reason to take it for granted that this scenario can usefully be played backwards and that now scientists will be able to identify a gene's critical function, or functions, when they have located, isolated, and sequenced that gene. "Predictive genetics" may work in a few special situations in which a particular DNA sequence points to specific and special characteristics that occur in only a few proteins, but most DNA sequences will not be that informative.

Geneticization

It has become fashionable to look for genetic explanations for health and illness. The argument runs like this: Environmental factors influence many aspects of our health, but despite the fact that people who smoke are at a greater risk of getting lung cancer than those who do not, not every one who smokes gets lung cancer. Conversely, not every one who gets lung cancer smokes. So, something other than smoking distinguishes people who get lung cancer from those who do not. To scientists who consider genes to be the basis of our entire biology, genes are the likeliest culprits.

As the human genome is analyzed at a new level of detail, correlations inevitably will turn up between certain DNA sequences and particular diseases or other traits. But, until the DNA sequences of large numbers of people have been looked at, it will be impossible to distinguish significant correlations from accidental ones. Unfortunately, at this point each correlation that results in a scientific paper tends to give rise to a news headline. When later scientific papers show the correlation to be false, that sometimes rates another headline, but often it does not.

Already the confusion is enormous. Within the last few years, genes have been announced "for" manic-depression, schizophrenia, alcoholism, and smoking-related lung cancer. The claims about manic-depression and schizophrenia genes were with-

159

drawn soon after their announcement and the gene for alcoholism met the same fate later, although another one has since crept into the news. These supposed identifications are invariably obtained with small numbers of people, and much publicity accompanies every such "discovery." Although, like mirages, many of these genes disappear when one tries to look at them closely, a confusion of claims and counterclaims is inevitable, and there are so many stories that people are left with the impression that our genes control everything. . . .

Exploiting the Findings

The sequencing of the human genome will leave us with a deluge of undigested genetic data. . . .

With such a bounty of unexplored data on the human genome available, there may be a temptation to hastily assign too great a causal role to many freshly mapped genes—simply because of their preliminary statistical associations with perplexing health problems.

Some people are bound to eagerly exploit such findings by publicly proclaiming that they offer 'scientific solutions' for 'fixing' everything from alcoholism and mental illness to homosexuality and learned disabilities.

David Suzuki and Peter Knudtson, *Genethics*, 1989.

In addition to the relatively few and rare conditions whose patterns of inheritance can be described by Mendel's laws, scientists and physicians increasingly speak of inherited "tendencies" or "predispositions" to develop more complex and prevalent conditions. In most of these cases, they are just using the word "gene" as shorthand for their belief that the condition is inherited biologically, even though they cannot be sure that it is and cannot predict who will inherit it. Complex conditions are variable and unpredictable, and involve a wide range of biological and environmental factors. It is not clear that identifying genes will give us a better picture of what is going on.

Considering the variety of social and economic risks all of us face, it seems a distraction from our obvious, daily problems to focus on the risks we may harbor in our genes. Worse yet is the implication that it would be irresponsible to go on living without this knowledge, even though there is little we can do once we have it. Yes, we can eat more healthful foods, but only if we can afford to buy them. And yes, we can decide to stop smoking and drinking, but only if the circumstances of our lives make such changes possible. Anyway, these changes would be good for all of us, irrespective of our genetic "predispositions." The unwarranted individualization of responsibility for our own health and that of our children and the fatalism genetic tests can engender may, in fact, prevent some of us from doing things we might otherwise do to stay healthy.

Unfulfilled Promises

In a 1989 article, published in the scientific journal *Genome*, the medical geneticist Arno Motulsky promises that "definite prediction of somatic [that is, physical] and some psychiatric disease will be increasingly possible in the future." Yet, in the very next sentence he points out that "in many conditions predictions will not be 100 percent accurate," but will only mean "that a given disease will occur with a greater statistical likelihood than expected in the general population." This is an odd kind of "definite prediction." It does precisely what [political scientist] Sylvia Tesh suggests scientists do when they try to appropriate our health: It promises major benefits that in reality add up to little.

Despite the scientific problems that surround the identification of genes "for" specific conditions, and the social and personal problems such predictions can entail, our current infatuation with genetics pushes genes into the foreground. Both the scientific pronouncements and the ways they are reported in the press often imply that, with a snap of their fingers, scientists will progress from the point at which they have identified a gene they suspect may be associated with some devastating condition such as cancer, to predicting whether an individual will develop the condition and, better yet, to curing or preventing it.

Let us look at an example: In September 1990 an Associated Press story announced that "Scientists . . . have cloned a gene that helps brain cells communicate, a step that may lead to improved drugs for schizophrenia and . . . may someday help doctors diagnose schizophrenia and Parkinson's disease before symptoms appear." This is the sort of thing scientists and science writers say to stimulate interest. We have all heard of schizophrenia and Parkinson's disease, and cloning "a gene that helps brain cells communicate" sounds impressive. The claim may be true, but the reality is much more complicated than they suggest. Scientists have identified a DNA sequence, implicated in the synthesis of a protein called the *dopamine receptor*, which occurs in brain cells. Dopamine is one of several small molecules that are released by some nerve cells in the brain and taken up by others. This is what is meant by the word "communicate." Scientists do not know what these brain cells say to each other, nor exactly how they say whatever it is that they do communicate. They just know that dopamine and other neuro-transmitters are involved.

Reductionism

Dopamine is also thought to be involved in some way in Parkinson's disease, since the tremors and other symptoms of some, though not all, people who have this condition are reduced when they take dopamine or compounds chemically related to it. Again, why it works is not understood, though the assumption is that, by binding and releasing dopamine, dopamine receptors may modulate the concentration and activity of this chemical in the brain. If that is so, the gene that specifies the receptor's structure may affect dopamine activity. The Associated Press story quotes scientists who suggest that once this gene has been isolated, they may be able to study the dopamine receptor in greater detail than was possible before, and that they may then be able to develop drugs that modify the receptor's interactions with dopamine.

In other words, scientists intend to use the gene as a biochemical tool to study the metabolism of dopamine and try to develop drugs that mimic or counteract its action. This is a reasonable plan of biochemical experimentation, but what makes it interesting to the public is its association with familiar diseases and stories about how the brain works. Molecular biologists are also attracted by these associations. They like to feel that they are getting at the root of human thought and action. We are back to reductionism: Brain function gets explained in terms of the activity of molecules in the brain, hence of the genes that participate in the synthesis of these molecules. Then, how people act gets explained in terms of brain function, hence of such molecules and genes.

We need to understand the patterns that underlie the grandiose scientific announcements and the ways they are reported in the press, because more and more of them are being made as it becomes technically easier for scientists to isolate genes and produce them in quantity. A host of therapeutic claims has been spawned by the idea that identifying a DNA sequence and the protein whose composition it specifies will lead to cures for a condition associated with that protein. Whenever a DNA sequence is isolated that specifies the composition of a protein involved with the ability of cells to multiply or stick together, scientists say they are on the road to curing cancer. Locating a gene that specifies a protein involved with cholesterol metabolism puts them on the road to conquering high blood pressure, strokes, and heart disease. And so on. But metabolic relationships and their derangements are too complex to permit such simplistic solutions.

There is an even more pressing argument against research

that seeks to identify genes "for" this or that condition. The development of tests to detect genes, or substances whose metabolism they affect, opens the door for the invention of an unlimited number of new disabilities and diseases. For any metabolite or other trait that has a normal distribution in the population, some people can be defined as having "too much" and others "not enough." (In mathematical terms, *normal distribution* simply means that most people cluster around some average value that gradually falls off toward zero on both sides of that average or *mean.* "Normal" in the colloquial sense means whatever the society wants it to mean.) Pharmaceutical companies and physicians stand to make a good deal of money from inventing new diseases as fast as new diagnostic tools are developed that can spot or predict their occurrence.

"Treating" Healthy People

Let us look at an example. Genentech, one of the first generation of biotechnology firms, markets a genetically engineered form of human growth hormone. This hormone previously could be obtained only in minute amounts, by isolating it from the pituitary glands of human cadavers. When the supply was limited, human growth hormone was only used to treat children with *pituitary dwarfism,* which results from the reduced secretion of this hormone by the pituitary gland. Once the hormone became available in quantity, physicians began to prescribe it to treat people who secrete normal amounts of growth hormone.

In one series of experiments, growth hormone was given to growing boys deemed "too short" for their age. A *New York Times Magazine* cover story on these experiments reports that Genentech scientists have suggested that it is proper to consider any child whose height falls within the lowest 3 percent of the population as suitable for treatment. But it is in the nature of characteristics like height that, no matter what their average distribution may be, there will always be a lowest—and highest—3, or 5, or 10 percent. Physician John Lantos and his colleagues point out that "of the three million children born in the U.S. annually, 90,000 will, by definition, be below the third percentile for height." This "treatment" is not without risks. There is no telling how the health of these children will be affected by daily injections of growth hormone. However, since growth hormone treatment costs about $20,000 a year per child, if each of these children received a five-year course of treatment this would constitute a potential market of about nine billion dollars a year for Genentech.

Height is not the only characteristic for which people are using growth hormone. Recently rumors have been circulating that athletes are using it to build up their muscles. Since the level of growth hormone varies from person to person, artificial supplements of it would be harder to detect than the metabolic steroids some athletes have used for this purpose. But human growth hormone, in excess, is by no means harmless. People whose pituitary gland secretes too much growth hormone often develop *acromegaly*, a condition that involves an overgrowth of the bones of the hands, feet, and face. Thus the use of this hormone to "treat" healthy people seems hardly justified.

Aging as Disease

Researchers have also suggested that administering growth hormone to old people could slow the aging process. A report of the use of synthetic human growth hormone for this purpose appeared in July 1990. The experiment, published by ten physicians in Milwaukee and Chicago, involved twenty-one men between sixty-one and eighty-one years old. These men reported no symptoms and were selected as subjects merely because, on two successive measurements, their hormone levels were in the lowest third of the normal range. Twelve of them were given sufficient amounts of human growth hormone to bring their levels into the range found in "healthy young adults." The other nine served as "controls."

Since all these men were healthy to begin with, the benefits of the treatment were measured by the following "symptoms": mass of fatty tissue, which tends to increase with age; overall muscle mass, which tends to decrease; and skin thickness and bone density, both of which tend to diminish. The experiment showed that, at a cost of about $14,000 a year, these indices could be brought into a more "youthful" range. However, the author of an accompanying editorial points out that long-term administration of growth hormone can elicit diabetes, arthritis, hypertension, edema, and congestive heart failure. Perhaps a more fundamental question is whether the fact that human growth hormone can now be produced in quantity justifies turning the normal process of aging into a disease.

Stories like these demonstrate that in a capitalist economy it is virtually impossible to develop products that can benefit only a few people. Once such a product becomes available, and especially if it has been expensive to produce, the producers will do what they can to expand the market for it, even when its wide use poses known dangers.

The market for artificial "improvements" will simply depend on where one decides to draw the line for what is to be labeled "abnormal." This is true for any numerical trait—height, weight, amount of body fat, metabolic rate, and so on. Now that biotechnology companies are producing growth hormone, an obvious next step is to produce an "antigrowth hormone" that promises to slow growth. Perhaps the companies could market it to parents of children, especially girls, who are predicted to be among the tallest 3 percent of the population.

If a boy is "too short" or a girl "too tall," if a woman's breasts are "too large" or "too small," if a man wishes he had been born a woman, or a woman that she were a man, they need only find a physician who can administer the right substance and their troubles will be over. Except that their troubles—and ours—will have just begun. There will always be people who would like to change their children or themselves and novel medical treatments won't cure such insecurities. As long as every deviation from the standard, prepackaged norm is considered "abnormal," physicians, geneticists, and the biotechnology companies will not run out of customers.

"*We have already made the transition . . . to the era when gene therapy is a reality.*"

Gene Therapy Is Beneficial

Thomas F. Lee

Thomas F. Lee is a biologist, teacher, researcher, and the author of the books *The Human Genome Project: Cracking the Genetic Code of Life* and *Gene Future: The Promise and Perils of the New Biology*, from which the following viewpoint is excerpted. In the viewpoint, Lee describes how gene therapy is increasingly being used to treat such devastating diseases as cancer, cystic fibrosis, and AIDS. Genetic engineering, Lee concludes, may help humans conquer many such diseases.

As you read, consider the following questions:

1. Describe the four forms of genetic manipulation, as outlined by the author.
2. What is cystic fibrosis, and how might gene therapy be used to treat it, according to Lee?
3. What does the public think about gene therapy, according to Lee?

As early as the 1960s the intriguing realization began to dawn that the exciting new developments in the technology for isolating and cloning genes might ultimately lead to their use as a means of combating disease. In 1967, Marshall Nirenberg, who was to receive a Nobel prize for his critical contribution to deciphering the language of the genetic code, predicted:

> My guess is that cells will be programmed with synthetic messages within 25 years. . . . Man may be able to program his own cells long before he will be able to assess adequately the long-term consequences of such alterations. . . .

It had been a long and convoluted path from Nirenberg's uncannily prescient remarks and the first approved gene therapy clinical trial almost exactly 25 years later. The tale of the journey along that path is one of science and medicine combining forces in a highly technical, sometimes controversial effort to combat genetic disease.

Approximately 14 percent of newborn infants are afflicted with some form of inherited physical or mental problem. These include chromosomal abnormalities such as Down syndrome, over 3,000 single gene disorders like sickle cell anemia or cystic fibrosis, as well as problems caused by little-understood interactions between genes and the environment. The latter group includes the common conditions of spina bifida and juvenile onset diabetes.

The Genetic Lottery

With very few exceptions, particularly in the case of the monogenic (single-gene) disorders, modern medicine has been able to offer only treatment of the symptoms with no hope of a cure. More than half of all monogenic diseases lead to an early death. Those people who survive beyond infancy often face a lifelong struggle with the limitations imposed on them by their genetic inheritance. The genetic lottery which mingles genes from two parents into a new unique combination in their child has been beyond the reach of medicine. The genes have always spelled out our potential, which even under optimal conditions is inherently circumscribed by the quality of the messages in our DNA.

Are we always to be prisoners of our genes? Can we do more to alleviate this human suffering than the dietary, pharmacologic, and surgical interventions, which offer some relief in varying degrees to some of those who are afflicted with genetic diseases? Beyond these, might we some day be able to prevent such genetic mishaps from occurring? Ultimately could we (and should we) develop enough control over genes so that we might remodel human beings into people who are free from all disease?

Consider the problem. Deep in almost all cells of the human

body there is a control center, the nucleus. Hidden behind the membrane surrounding each nucleus are 46 chromosomes, long coiled strands of DNA spelling out the instructions in the 100,000 or so genes scattered along their twisted length. The actual functional genes make up only about 2 percent of the chromosome, while the rest is a puzzling patchwork of DNA sequences of as yet unassessed utility.

At any one time only a small fraction of those functional 2 percent is at work coding for the cell's proteins. The rest is turned off by a little-understood complex control system within the cells, a feedback system which sees to it that certain genes act only under specific circumstances. This means that in the retina of the eye, for example, a unique set of genes is operating which is dormant in the liver or the skin. This results in the variety of form and function necessary in the cells, tissues, and organs making up the complex human organism.

From a Distant Ideal to Reality

The microscopic genes are inaccessible to any kind of surgical intervention. Defective genes cannot be physically removed from the living cells and replaced by normal ones. The genes are there to stay. But could it ever be possible to ameliorate the effects of the defective genes not by symptomatic treatment of the entire body but by adding normal genes to the nuclei of at least some of the body's cells?

That is precisely what is now being done. Gene therapy is now in its early stages of clinical trials and development. In the space of a few years it has passed from being a distant ideal in the minds of a few pioneers to a reality. It is not, however, the only conceivable form of genetic manipulation in humans. There are four such possibilities:

1. *Somatic-cell gene therapy* aims at introducing genes into some of the somatic (body) cells to correct a genetic defect. This is the type of manipulation most commonly referred to as "gene therapy." In this case any effects would be limited to the person involved. The new genes would not be passed on to future generations.

2. *Germ-line gene therapy* would result in the correction of the genetic problem in an individual's reproductive cells so that it would no longer be passed on to his or her offspring. If performed early in the individual's life, perhaps even at the embryonic stage, the disorder might be corrected in the person being treated as well.

3. *Enhancement genetic engineering* would consist of inserting a gene at a pertinent stage in a person's development so that a specific characteristic would be improved or enhanced, such as the introduction of a gene to increase the amount of

growth hormone production, resulting in a taller individual.

4. *Eugenic genetic engineering* might come about by inserting genes to change or improve complex traits which arise from the interaction of many genes with the environment, such as personality or intelligence. If either the eugenic or enhancement engineering were done at an early enough stage, such changed traits might be passed on to future generations.

By 1993 over 40 clinical experiments for gene insertion into humans, most in the United States, had been approved by an unprecedented series of intensive regulatory reviews. The diseases to be confronted include various forms of cancer, liver failure, hemophilia, high blood cholesterol, and even AIDS.

Fighting Diseases

The field of "gene therapy" has spawned dozens of experiments aimed at treating ailments ranging from cystic fibrosis to brain tumors. The goal is to transplant new genes into humans to do the work of defective ones—or to give patients extra genes useful in fighting diseases.

Time, December 13, 1993.

Enhancement genetic engineering and germ-line genetic manipulation are being carried out in animals. These experiments have given scientists the technical ability to apply some of these genetic modifications to human subjects. Eugenic genetic engineering, for now at least, is beyond our capabilities. . . .

We have only just begun to use genes in our battle against cancer and other human diseases. Bear in mind, human gene therapy is still in its very early stages. . . . However we can anticipate what may develop into the twenty-first century by briefly reviewing clues that scientists already have which may lead to tantalizing possibilities.

Genes Versus Cancer

Experts predict that the major emphasis in human gene therapy experimentation over the next few years will be directed at the treatment of cancer. Emphasis will be on genetically modifying cells of the body's own immune system to fight tumors. In this way, the same genes may be used to treat many different forms of cancer. . . .

There is a pressing need for innovative treatments against this devastating disease. Each year more Americans succumb to cancer than died in World War II and the Vietnam War combined. . . .

As 1992 grew to a close, the International Conference on Gene

Therapy of Cancer was held in San Diego, California. The proceedings revealed a remarkable burst of activity in research on gene-therapeutic techniques, as well as reports on success in identifying genes that either inhibit cells' ability to mutate into tumor cells or cause tumor cells to revert to normal ones. By early 1993 over a dozen clinical trials had been approved for gene therapy aimed directly at the cure of various cancers, including ovarian, melanoma, brain, kidney, and lung cancers.

Despite the fact that there are still many technical problems to overcome before gene therapy for cancer can become standard treatment, if research and testing continue at the current rapid rate, the decade of the 90s may perhaps see genes become a powerful tool in the fight against cancer, one of humanity's most formidable enemies.

Cystic Fibrosis

Cystic fibrosis is among the most commonly inherited disorders among Caucasians. There are now about 30,000 people with CF in the United States. Children who inherit two copies of the faulty gene, one from each parent, are not able to excrete chloride from the moist cells lining the lungs, sweat glands, intestines, and pancreas. Thick mucus builds up, particularly in the lungs, breeding numerous infections. Few survive beyond their twenties.

In a triumph of molecular biology, Francis Collins, Lap-Chee Tsui, Jack Riordan, and their co-workers ended a 10-year quest in August 1989. They had tracked the CF gene to its home on chromosome 7. Soon several groups had managed to put the gene into lung cells isolated from CF patients, resulting in the restoration of normal chloride transport across the cell membranes. Then Ronald Crystal at NIH [National Institutes of Health] injected the lungs of rats with a nasal drip of adenoviruses (cold-causing viruses) that had been stripped of their infective genes and replaced with normal genes. The rat lung cells produced the normal human gene product, CFTR, for up to six weeks.

In December 1992 the NIH approved three proposals to use genetically altered cold viruses into the lungs of 25 people suffering from cystic fibrosis. . . .

While gene therapy trials to combat cystic fibrosis are just getting under way, officials at the Cystic Fibrosis Foundation are expressing cautious optimism for the future of CF gene-based treatments for this devastating disease.

Muscular Dystrophy

One of every 3,500 boys is born without a functional gene for producing the important muscle protein dystrophin. This deficiency leads to muscular dystrophy, a condition in which the muscles gradually waste away. Boys born with Duchenne muscular dystrophy, the most common childhood form of the disorder, usually die in their early 20s. There is no known cure. . . .

In MD the genetic defect is seemingly inaccessible, hiding deep within the long, fibrous cells of the skeletal muscles. . . .

In 1990 Peter Law at the University of Tennessee in Memphis had pioneered an ingenious approach to a potential MD treatment. He injected the big toes of several young MD patients with immature muscle cells taken from either their fathers or their brothers. The small fragments of donated muscle had been first cultured in the laboratory. Within days, millions of small, elongated myoblasts—repair cells that can rebuild injured muscle—grew from the small reservoir of myoblasts always present. Those myoblasts, when injected into the toe muscles, fused with them and began to produce some dystrophin.

Later in that same year, two research teams, one led by Eliav Barr and Jeffrey Leiden of the Howard Hughes Medical Institute and the other by Helen Blau at the Stanford University School of Medicine, took advantage of myoblasts in developing another ingenious way to ferry genes into muscles. They put the gene for human growth hormone into myoblasts and injected these into mouse muscles, resulting in significant blood levels of the hormone for up to three months.

Will the future bring news that the dystrophin gene can be insinuated into muscles by myoblasts or injection or other means in enough quantity so that once-useless muscles can function again? Or is more than just dystrophin needed? Researchers predict that myoblast gene therapy may have an equally promising future as a delivery method for many other therapeutic genes. Given methods for controlling the rate of activity of newly introduced genes, not an easy challenge, we will perhaps be able to fashion myoblasts which when implanted will not only create dystrophin where needed but release insulin to control diabetes or generate Factor VIII, the blood-clotting protein missing in one type of hemophilia.

The Heart

The cholesterol in our blood is a mixed blessing. It is a vital chemical which makes up part of the membrane of every cell in the body. Unregulated, it can lead to an early death. The concentration of cholesterol in the blood is the result of a balance struck by our dietary intake and cellular metabolism. A crucial element of that control is the LDL receptor, a protein which pokes through the surface of liver cells and snares passing molecules of low density lipoprotein (LDL), often referred to as the "bad" form of cholesterol. Transported inside the cell, the

LDL is digested and removed.

One in 500 people inherit one abnormal gene for making LDL receptors, causing a mild form of familial hypercholesterolemia, while 1 out of a million are born with a severe case of this genetic disease. In either situation, blood cholesterol levels are high, in the latter instance so high that the person develops severe coronary artery disease and rarely survives beyond the second decade of life.

In a few cases people have been helped by the drastic procedure of a liver transplant. Now, there may be a way to bolster the LDL filtering capacity of the genetically deficient liver through gene therapy. James Wilson and co-workers at the University of Michigan Medical Center have used retroviruses to place the normal LDL receptor gene into liver cells removed from rabbits with hereditary high cholesterol. Transplanting the genetically corrected cells into the donor rabbits' livers resulted in a 30–40 percent decrease in their blood cholesterol. . . .

Others . . . have placed genes in animal arteries by implanting vascular grafts seeded with genetically modified cells. In the future we may see such grafts used to deliver a wide range of therapeutic genes, such as the gene for tPA, a naturally occurring chemical which aids in dissolving blood clots. Many other conditions needing gene replacement might be targeted as well besides those that are implicated in cardiovascular disease. As usual, a major stumbling block will be to develop gene constructs which will put enough gene product into the blood at safe, useful concentrations.

AIDS

Paradoxically, gene therapy researchers are calling on the deadly AIDS virus HIV to carry resistance into T cells, the very cells in the human immune system which the virus targets for destruction. The scientists first remove the infectious genes (in the form of RNA in the retroviral HIV) and replace them with genes designed to block HIV replication. Trials will soon be under way in which blood will be taken from HIV-infected patients. T cells that remain capable of identifying and killing other cells harboring the virus will be carefully weeded out of the sample and cultured in large numbers. The hope is that these, when returned to the patient, will seek out and destroy infected cells.

Other ways of tricking infected cells into harming the virus are in the planning stages. We may someday be able to target genes to infected cells where the genes would trigger cell suicide. Another possibility might be to stimulate many cells to make CD4, the cell surface protein recognized by the AIDS virus, thus luring it to attach itself to the decoy CD4 rather than that of real

T cells. This would spare the vital T cells from destruction and leave the virus clinging to CD4 and thus unable to do further harm. French Anderson is collaborating with Dr. Robert C. Gallo of the National Cancer Institute on just such an effort. . . .

In the short run, making better use of existing drugs and searching for new ones will continue to be a major priority. But in the background, gene therapy remains an ideal whose time may come. . . .

Public Opinion

What do we know of the attitudes of the general public about gene therapy? Many people were surprised by a Harris poll conducted for the March of Dimes in April of 1992. It was perhaps not remarkable that 79 percent of the respondents said that they would undergo gene therapy to correct a serious or fatal genetic disease, or that 73 percent thought that potential dangers from genetic alteration of cells is so great that strict regulations are needed.

What surprised many interested observers was that 43 percent would approve of using gene treatments to "improve the physical characteristics that children would inherit," and 42 percent would support it to "improve the intelligence level that children would inherit." This high level of confidence was expressed despite the fact that 87 percent admitted to knowing little or nothing about gene therapy itself. March of Dimes officials cautioned that there is an obvious need to stress that the goal of gene therapy is "to make sick babies healthy, not normal babies perfect.". . .

The poll indicates that many people already approve of trying to engineer our genomes to enhance our memory, intelligence, or immune system. Might it even someday be considered an obligation to use these tools to free us and our descendants from unwanted genetic limitations?

On the wall in French Anderson's meeting room when he was a genetic researcher at NIH was a framed quotation of a few lines from *Hamlet*:

Diseases desperate grown
By desperate appliance are reliev'd
Or not at all.

Surely many currently afflicted with intractable diseases desperately hope that science will find a cure for them. We have already made the transition from the recent past, when the notion of using genes as a medical tool was widely held as a "desperate appliance," to the era when gene therapy is a reality. We will soon know whether such therapy will advance to the point where the hopes of those with genetic disease will be realized.

"It is ridiculous to reduce a condition to a singular hereditary notion of genes."

The Benefits of Gene Therapy Are Exaggerated

Pat Spallone

Gene therapy has been heralded by many as the ultimate solution to such serious inherited diseases as cystic fibrosis and Huntington's disease. But in the following viewpoint, Pat Spallone argues that gene therapy poses social, ethical, and medical problems. She fears that treating all inherited conditions as "diseases" could result in the stigmatizing of affected individuals and in harmful attempts to control human reproduction. Spallone is a freelance researcher and writer associated with the Centre for Women's Studies at the University of York in Great Britain. A former biochemist, she is the author of the books *Beyond Conception: The New Politics of Reproduction* and *Generation Games: Genetic Engineering and the Future of Our Lives*, from which this viewpoint is excerpted.

As you read, consider the following questions:

1. What is the XYY syndrome, and how does Spallone use it to illustrate her opposition to gene therapy?
2. What are the "eugenicist ideals" within genetic medicine, according to Spallone?
3. What happened in the Mediterranean village of Orchemenos, and how is it an example of genetic control gone awry, in the author's opinion?

*What happens when several genetic factors contributing to suscep-
tibility to cardiovascular disease have been identified, a question
now the focus of a great deal of research? It should then be possible
to categorize groups of people who are at risk, and who could (and
should) take preventative steps, but only by some complicated population screening.*

Peter Newmark, *Nature*

*Shall we create a special class of genetic outlaws who come slouch-
ing through our streets, branded by yellow armbands and with bro-
ken double helices advertising their defective status?*

Nancy Wexler, *Genetics and the Law II*

Once upon a time, the story goes, before there was gene splicing and gene cloning, there were only three ways to study inherited disease, or what was thought might be inherited:

- clinicians and researchers could study inheritance patterns
  through generations of a family by observing which members were affected
- they could take blood samples to diagnose conditions for
  which biochemical tests were available, in order to identify
  persons affected or persons who carry the condition
- they could look under a microscope at cells and chromosomes to look for abnormalities or what they thought were
  abnormalities

(I take pains to qualify what observers were and are doing. The
history of attempts to correlate human characteristics with
heredity is riddled with dubious undertakings. For example, an
XYY chromosome pattern was linked to delinquency, psy-
chopathology and criminal behaviour in men in the late 1960s
and early 1970s. This claim is now generally discounted.)

Genetic engineering changed the scope of all that. Now it is
possible to isolate genes (pieces of DNA). It is possible to clone
them and study them. Once a gene is associated with a condition, it is possible to use a 'gene probe' to 'search' an individual's DNA to see if they carry the gene in question.

Genetic engineering is the foundation of the so-called 'new genetics in clinical practice'. The 'new genetics' are direct gene
analysis and direct gene intervention; the 'old' genetics is based
on less direct methods of gene analysis: for example, sickle cell
anaemia may be identified by looking at a person's blood cells
under a microscope. The 'new' genetics is most interested in
studying the fine structure of genes and how they work. . . .

Desperately Seeking Genes

There is an incredible amount of genetics research geared towards identifying 'disease-causing' genes and genetic mechanisms associated with diseases. This research is not limited to
'classical' hereditary conditions such as sickle cell anaemia,

haemophilia or cystic fibrosis. It includes many other types of illnesses and conditions that are not illnesses. . . .

Questions of research priorities and context arise here. Genetic explanations may be the scientific fashion, but as a fashion they diminish those other ways of understanding health and ill health, and appropriate medical care. Although few medical scientists would deny the need to look at the whole context of an illness, including environmental and social factors, the bulk of genetic research does not take these into consideration. Instead the approach is focused at the molecular gene level; then, after a genetic link is 'found', so a cure or therapy will become obvious. The hunt for genetic causes tends to overshadow the substantial evidence regarding other factors in health and illness. If a person's individual biology becomes the major focus of genetic explanations and genetic cures, what are we losing here? Equally important, what do we get from these quests?

The Burden of Genes

Medical students and students of genetics may learn that genetic disease is a 'burden on the community', that the spread of 'morbid genes' burdens the population. Thus any medical intervention which aims to decrease the genetic burden, for instance through genetic counselling, gene screening and selective abortion, is beneficial to the population as a whole, as well as to the individual woman and family involved. This idea of removing 'disease genes' from the population, although not the sole aim of genetic medicine today, remains acceptable among some medical professionals, including some human geneticists.

No such idea is inevitable, so where did this one come from? . . .

According to historian Daniel J. Kevles in *In the Name of Eugenics: Genetics and the Uses of Human Heredity*, genetic screening was an idea which initially came from the eugenics movement, a social movement which was dedicated to the idea of differences in biological 'quality' among individuals. Among its proposed aims was human selective breeding for the 'more suitable' human qualities, through education or compulsory laws or both.

Eugenics always had a close association with the science of genetics. . . . Social reformers who embraced eugenics believed that the empirical findings of genetics could be put to use to breed an innately 'better' human population. . . .

Eugenics reached a horrifying depth as the basis of the political philosophy of the Nazi 'racial hygiene' or 'genetic health' programme which began as an expulsion or internment of Jews, Poles, gypsies, homosexuals and anyone else deemed inferior, and which ended in mass executions and plans for genocide. To maximise reproduction among those considered of superior and

'Aryan' stock, breeding programmes were set up, with an accompanying ideology of motherhood that elaborated the role of 'Aryan' women as patriotic breeders.

In the aftermath of the Nazi population programme, genetics lost its close attachment to eugenics, but the science of human genetics never completely lost its ties with eugenical principles of one kind or another. Now the bio-revolution gives the word eugenics a new lease of life. The word 'eugenics' is even making a come-back as an acceptable principle.

Proponents of the new eugenics of the new genetics say it is now going to be a good thing based on a good science; this time eugenics will not be oppressive, because it is not about race but about genes which can be observed and measured scientifically; it is not about eliminating certain types of people, but about dealing with genetic medical conditions in any number of ways, for example, creating better diagnostic tests and gene therapies. Indeed the objective of certain forms of gene therapy is not to eliminate 'morbid genes' from the population, but to cure conditions 'caused' by the genes; people with hereditary illness who in the past would not have grown up to reproduce, may in future do so. In fact some analysts would reject the term 'new eugenics', saying genetic intervention today is not eugenicist for these very reasons. All in all, the argument goes, knowledge of genetics and new gene technologies is highly scientific, and should work to improve the quality of life.

However, the genetics of the past also always promoted itself as highly scientific and based on biological facts, but was none the less influenced by eugenicist ideas, and formulated faulty correlations of heredity and behaviour such as the XYY syndrome. And since influential social movements do not disappear as swiftly as eugenics supposedly has, we should judge carefully the present science of medical genetics which proposes that if the presence or absence of a gene is the cause of an ailment, genetic technology may provide the solutions.

People and Genes

The language of medicine tends to hide the person behind an impersonal vocabulary of body parts, while scientific terms tend to mask the fact that when one is talking about a 'deleterious gene', one is saying something about a human being. In the parlance of genetic medicine we hear about 'genetic defect', 'defective gene', 'disease-causing genes', 'chromosomal defects', 'genetic abnormality', 'morbid genes', 'genetic burden', 'genetic risk', 'molecular pathology' and functions which are 'deviant' rather than 'normal', but it still comes down to people's characteristics, and social categories of normality and abnormality. It is easy to lose sight of this in the world of biochemicals, genes and body parts. And it remains that someone must decide which genes and conditions create and constitute a 'burden'. More to the point, someone must decide that genes *are* the problem, the burden. As for the implications: Oxford philosopher Jonathan Glover entitled his book on the subject of the new human genetics, *What Kind of People Should There Be?* . . .

Premises Begin to Surface

I am a child of a survivor of the Holocaust . . . and many millions of others suffered very directly and convincingly the results of policies based on dubious genetic traits. . . . While I don't see gas ovens being built in the United States, I do see that many of the premises which led to that terrible time are again beginning to surface here.

Jeanne Stellman in *Generation Games: Genetic Engineering and the Future for Our Lives*, 1992.

I am not implying that there is a monolithic eugenicist objective within medical science. The problem is not the existence or otherwise of a secret international conspiracy of eugenicists. The point is that there is a lack of recognition of social prejudices and what can only be called eugenicist ideals and aims within genetic medicine today—that people's lives may beneficially be controlled through genetic interventions; and that these are supporting far-flung theories of genes for conditions as diverse as Huntington's chorea, cystic fibrosis, schizophrenia and susceptibility to occupational illness.

Huntington's Chorea

Huntington's chorea was one of the first illnesses ostensibly to show 'a gene defect link of diseases of the mind'. It is a degenerative disorder of the nervous system; it was known to be inherited long before any genes became associated with it.

People who get Huntington's chorea are usually healthy until the onset of symptoms around the age of 40, when movement becomes uncoordinated and involuntary. They become progressively further mentally and physically debilitated. Huntington's chorea has been described in *Nature* as an 'inherited neuropsychiatric disorder of unknown etiology [origin]'; the 'disease shows itself as a gradual, irreversible deterioration in physical condition and personality,' according to *New Scientists*.

James Gusella of Harvard Medical School in Massachusetts led the team which isolated the 'gene marker' for Huntington's chorea by studying a community in Venezuela where there is a

178

high incidence of Huntington's chorea. A gene marker is a section of DNA closely linked to the gene being hunted. Various gene markers have now been isolated for Huntington's chorea by different researchers trying to locate the elusive 'disease gene'. . . .

The public were left with yet another promise that the discovery of the gene marker meant that Huntington's chorea could be eradicated (yes, eradicated) within a few generations. Reporting from a scientific conference, Germaine Greer had this to say:

> 'In one generation we could eliminate Huntington's chorea from the gene pool,' said one of the most distinguished gentlemen present. The only way this could be done would be by sampling genetic material from all foetuses conceived by people with the disease, and aborting all those that carried the marker for Huntington's. There was nothing in the great doctor's voice to indicate that he found this anything but a delightful prospect or that he anticipated any resistance from the Huntington's sufferers themselves. He had never heard of Woody Guthrie, who said, when he was in the final throes of the disease, that he thought he had had a good and useful life.

The ethical dilemma most commonly recognised is that for people at risk from Huntington's chorea because of their family background, the genetic test may give them their lives back, or it may give them a death sentence. There is no treatment available for Huntington's chorea, just one of a growing list of conditions for which presymptomatic screening tests are available. Greer observed that about all the new genetics can offer a person suffering from Huntington's is to wish they had never been born.

Some people who know that Huntington's chorea is in their family history say they want to know if they carry the gene; others say they do not want to know. The personal dilemma has attracted notice, as well it might, but it seems to have got stuck at a point where it leaves the individual responsible for choosing testing rather than opening up the discussion to consider how the focus on genes-the-cause changes medical attitudes about certain diseases; how a legal apparatus is emerging about genetics; the stigma that may accompany a person diagnosed with a genetic condition; the pressures on people to embrace the gene technology solution, to name a few pressing problems. . . .

The Tragedy of Orchemenos

Orchemenos is a Mediterranean village which became the setting for a medical experiment that went terribly wrong. In Orchemenos, there was a high incidence of sickle cell anaemia, a heritable illness which may be passed on to a daughter or son if both biological parents are 'carriers' of the disease. . . .

In Orchemenos, the medical teams estimated that one in four of the villagers had sickle cell trait, that is, they were carriers of sickle cell anaemia. 'Scientists decided to help,' explained science writer Robin McKie in relating the story. They decided to do so through the village marriage arrangement customs, to persuade carriers to marry noncarriers, and thus eventually free the village of sickle cell anaemia. 'Villagers were screened, carriers warned and the scientists departed.' The débâcle began. Carriers were shunned; they ended up marrying each other and the rates of sickle cell in newborn babies rose.

McKie concluded that the problem was education and counselling 'about the realities of genetics'. Otherwise, 'people had no way of making sense of medical advice.' He suggested that the lesson it held for Britain if it were to avoid a similar horror with new developments in gene technology was that, 'The Government must find ways of institutionalising genetic counselling throughout the country's general medical practices, family planning clinics and classrooms as a matter of urgency.'

I learned different lessons. This *was* a reality of genetics, and another example of the huge deficit of human understanding in the scientific 'solutions' of gene technology, and its potentially disastrous consequences. No matter how good-willed an individual doctor or scientist is in their wish to alleviate the suffering of human beings, this particular experiment was at best ill-considered, at worst a paternalistic attempt at scientific population planning. The implementation of an institutionalised genetic control of reproduction based on eugenic values made human beings and lives invisible. Human needs were interpreted from the abstract 'laws' of genetics. Human needs were sacrificed to the abstract idea.

There is something here to suggest that institutionalisation does not prevent people being stigmatising, but normalises it. What power or choice does an individual have in accepting or rejecting a genetic therapy if that is the only help or offer or when it is perceived as the best alternative? At the moment, many women who undergo genetic screening do decide to carry on with their pregnancies when cystic fibrosis is foretold by screening. For everyone who lives with it, the pain of having cystic fibrosis or sickle cell anaemia is not the only thing their lives are about.

Huntington's chorea, cystic fibrosis and sickle cell anaemia may well be illnesses which must be understood as inherited—but not *only* as inherited. First, it is ridiculous to reduce a condition to a singular hereditary notion of genes, not least considering that scientists admittedly have only the haziest ideas about how genes work in the body. Heredity is not simply the sum of one's genes. Hereditary does not simply mean genetic. Anyway, even if scientists 'knew more', it is no excuse for establishing any eugenicist policy. It does not mean medical care should inevitably become a policy of screening to eliminate a type of person.

Periodical Bibliography

The following articles have been selected to supplement the diverse views presented in this chapter.

Natalie Angier	"DNA Sleuths Follow Disease's Tracks to Two Very Different Genes," *The New York Times*, January 17, 1995.
Andre Bacard	"Sexoids: Better Sex Through Genetic Engineering?" *The Humanist*, May/June 1993.
David Beers	"The Gene Screen," *Vogue*, June 1990.
Jared Diamond	"The Cruel Logic of Our Genes," *Discover*, November 1989.
The Economist	"Engineering Health," March 19, 1994.
The Economist	"Hippocrates' Dilemma," March 19, 1994.
William Friend	"Frontiers of Genetic Research: Science and Religion," *Origins*, January 9, 1995. Available from 3211 4th St. NE, Washington, DC 20017.
Denise Grady	"The Ticking of a Time Bomb in the Genes," *Discover*, June 1987.
Joan O.C. Hamilton and Naomi Freundlich	"The Genetic Age," *Business Week*, May 28, 1990.
Hastings Center Report	"What Research? Which Embryos?" January/February 1995. Available from 255 Elm Rd., Briarcliff Manor, NY 10510.
Andrew Kott	"Can Genetic Testing Help Prevent Illness?" *American Medical News*, August 22–29, 1994. Available from 515 N. State St., Chicago, IL 60610.
Geoffrey Montgomery	"The Ultimate Medicine," *Discover*, March 1990.
Richard John Neuhaus	"Don't Cross This Threshold," *The Wall Street Journal*, October 27, 1994.
Joseph Palca	"Doing Things with Embryos," *Hastings Center Report*, January/February 1995.
Peter Radetsky	"The Mother of All Blood Cells," *Discover*, March 1995.
Philip E. Ross	"Rebuilding Bone with Less Pain," *Forbes*, February 13, 1995.
Inder M. Verna	"Gene Therapy," *Scientific American*, November 1990.

How Should Genetic Engineering Be Regulated?

Chapter Preface

The U.S. government regulates many industries to protect the health, safety, and property of citizens. For example, the Food and Drug Administration regulates the safety of drugs, cosmetics, medical devices, foods, food additives, and other products. While some view regulation as a necessity, others believe it interferes with the growth of business and industry and does little to protect citizens.

Attitudes about the degree of regulation an industry requires often depend on the perceived dangers of that industry. Genetic engineering is no exception. Those who believe genetic engineering poses serious threats to humans and the environment are more likely to advocate strict regulation of the technology than those who contend it is safe. Michael W. Fox, the author of *Superpigs and Wondercorn: The Brave New World of Biotechnology and Where It All May Lead*, is among those who view genetic engineering as potentially risky. He writes, "Clearly genetic engineering is as much a Pandora's box as it is a cornucopia of wonderful possibilities. . . . Without congressional and state oversight and international coordination to minimize risks to the environment and to the very fabric of life itself, we could be on the threshold not of some biological utopia, but of our own nemesis."

Others contend that by overstating the risks posed by genetic engineering, critics have created and inflamed unjustified fears, which in turn may lead to overregulation of the technology. According to Karen Anne Goldman Herman, an attorney for the inspector general of the National Science Foundation, "Despite the abundance of data indicating the beneficial potential of biotechnology and the absence of harmful incidents, genetic engineering has aroused considerable public suspicion and from some quarters a demand for government oversight out of proportion to the demonstrated risks."

The authors in the following chapter examine the threats posed by genetic engineering and present their views concerning how much regulation, if any, this new industry requires.

"A more comprehensive and straightforward federal . . . regulatory scheme . . . could foster both the development of biotechnology and protection of human health and the environment."

Federal Regulation of Genetic Engineering Can Be Effective

Karen Goldman Herman

Exaggerated fears about the risks of genetically engineered foods could lead to excessive regulation of the biotechnology industry and prevent the development of beneficial new products, Karen Goldman Herman maintains in the following viewpoint. To keep this from happening, she advocates a federal policy that would ease consumers' fears about genetically engineered foods while permitting the industry to grow. Herman is an attorney in the Office of Inspector General at the National Science Foundation. She has a Ph.D. in neurobiology from the University of California at San Francisco, and was a postdoctoral scientist at the National Institutes of Health and the California Institute of Technology.

As you read, consider the following questions:

1. What are some of the risks posed by genetically engineered foods, in the author's opinion?
2. What errors do federal agencies often make concerning the regulation of new technologies, according to Herman?
3. What role should state governments have in regulating genetically engineered foods, in Herman's opinion?

From "Issues in the Regulation of Bioengineered Food" by Karen Goldman Herman (now Karen A. Goldman), *High Technology Law Journal*, vol. 7, no. 1 (1992). Reprinted by permission of the author.

It has been 20 years since the groundbreaking 1975 meeting at the California conference center Asilomar where scientists discussed the emerging technology of molecular biology, its vast potential and the possible risks that could result from the ability to transfer DNA from one organism to another. Since then, a number of biotechnology-derived pharmaceutical products have already gone on the market, and the first food and agricultural products have been approved or are close to approval. Many more such products are under development, and there have been no adverse impacts on human health or the environment. Rather, reputable scientific and medical sources stress the potential of biotechnology to improve human health and nutrition, and to ameliorate the adverse impacts of traditional agricultural practices on the environment. Despite the abundance of data indicating the beneficial potential of biotechnology and the absence of harmful incidents, genetic engineering has aroused considerable public suspicion and from some quarters a demand for government oversight out of proportion to the demonstrated risks. The negative perception and resulting regulatory response threatens to adversely affect the development and competitiveness of this fledgling industry, and may also delay or even block the introduction of beneficial products. . . .

Risks and Regulatory Issues

Although new and unknown technologies are often viewed with suspicion, some features of biotechnology make it particularly susceptible to an exaggerated perception of risk. Public concern may stem from scientists themselves, who initiated a moratorium on some aspects of genetic engineering in 1974. While scientists have since grown comfortable with the technology, the public perception of unreasonable risks lingers on. A survey by the Office of Technology Assessment found that 52% of the public "believes that genetically engineered products are at least somewhat likely to represent a serious danger to people or the environment." Biotechnology often suggests the "Frankenstein image." While the current technology generally changes but a single gene, producing a relatively small modification, many people may believe that any interspecies exchange of genetic information results in a dramatic change. Perhaps such views underlie the finding that 24% of a group aware of biotechnology felt that creation of hybrid plants and animals through genetic engineering is morally wrong. Another aspect of biotechnology that invites public concern is the ability of living things to reproduce; thus any deleterious effects of genetically engineered organisms have the ability to escape human control and self-perpetuate. This leads to a fear that although a deleterious result is unlikely, if it occurs, the outcome could be a problem of substantial magnitude. Such fears of an unlikely but potentially disastrous outcome could greatly hinder the progress of biotechnology. A majority of the public would object to the use of genetically engineered organisms if the risk were unknown.

Encouraging Development, Ensuring Safety

Biotechnology offers great potential in many areas, particularly in continuing improvement of our capacity to develop products that are of critical importance to the future of American agriculture. It is vital that products of biotechnology be regulated based on firm scientific principles and meet the same high standards of safety and efficacy as do those products made through conventional technology. Thus, we at the USDA [U.S. Department of Agriculture] seek to foster a regulatory climate that encourages innovation, development, and commercialization of beneficial new agricultural products derived from biotechnology, while implementing a responsible policy that limits potential of real risks.

Kenneth A. Gilles, testimony before the U.S. House of Representatives Science, Space, and Technology Subcommittee on Natural Resources, Agricultural Research, and Environment, May 5, 1988.

Food products of biotechnology generate their own specific concerns. Production of bioengineered food usually involves not only a consideration of the safety of the food for human consumption, but also the safety of environmental release of the altered plant. The public's perception of potential danger from food biotechnology is enhanced by its heightened awareness of environmental damage from the introduction of exotic species and of health problems that are manifested only decades after exposure to the causative agent. Yet many similar risks from food stem from traditional agricultural and plant breeding practices that are essential to provide sufficient food to the growing population or to assure the taste, quality, and convenience that consumers and farmers have come to expect. Thus, society accepts environmental risks of pesticide use and dispersal of domesticated plants and animals within certain limits and tolerates low levels of pesticide residues in food. Other risks from food are inherent in the food itself. Food contains many naturally occurring toxicants and carcinogens that are nearly unavoidable in the ordinary diet.

Considering the Risks

Biotechnology presents few risks beyond those already accepted in traditional foods. As to their environmental risks, the National Research Council states that "[c]rops modified by molecular and cellular methods should pose risks no different from those modified by classical genetic methods for similar traits." Bioengineered organisms' potential for dispersal and environmental disruption is generally similar to their traditional counterparts. Society has long accepted the fact that traditional plant and animal breeding practices may change the nutrient or toxicant levels in the food or alter an organism's potential for environmental dispersal. Although traditional methods usually enhance the safety of the food, they have occasionally increased the level of a deleterious component. The use of antibiotic resistance marker genes in the production of bioengineered food has raised some questions, but most experts agree that the genes should cause no health or safety problem. Bioengineering, as an extension of traditional breeding practices, should pose no greater concern over the safety of the food consumed; it should actually be safer since the recombinant techniques are more specific and thus less likely to produce unwanted side effects such as increased levels of toxicants or weediness. Indeed, as considered above, bioengineering may lower both the environmental and food consumption risks.

New technologies are particularly difficult to regulate when their risks are unknown, but to reap the benefits of such advances it is important that regulation be based on risk and not succumb to exaggerated perceptions of danger. Peter Huber has argued that regulation of new technologies by federal agencies often involves screening that eliminates small risks at the expense of lost opportunity costs of unknown magnitude. By contrast, old technologies are usually subject to more lenient standard-setting regulations. Thus, regulations often preserve the present level of safety by tacitly accepting risks posed by old technologies while excluding new technologies with potentially large benefits. The assumption behind screening new risks but setting standards to limit old risks is that barring new risks is economically and socially less costly, because both producers and a market for the new products are not yet established. The benefits of the new products are generally not considered in the screening process because their values are speculative.

A Comparative System

Thus, to foster technological advance and its resultant benefits, Huber argues that a comparative system of regulation of old and new risks, one that permits new technologies functionally similar to established technologies and of no greater risk, should be implemented. The comparative approach, allowing a new risk, is justified when the old, risky product is one that society accepts either because it is essential or desirable. As Huber states, "[E]xcessively strict regulation of the safer-than-average prod-

187

uct[s] will drive consumption toward the more hazardous ones." Comparative regulation, on the other hand, would favor the safer product, particularly because modern technology usually replaces an old outmoded source of risk rather than adding to it.

Huber suggests a four-step process for implementing comparative regulation:

1) The agency must define a risk market comprising products that are functional substitutes for each other.

2) It next must identify typically risky products already allowed to compete in that market.

3) The agency must then compare the risks of the new substitute with those of products not in fixed supply and already in the market. Only the less safe substitutes must be excluded or otherwise regulated.

4) If a new product offers exceptional price or other advantages over existing, more hazardous products, introduction of the safer product could conceivably increase net risk by increasing total consumption. As a final step in comparative regulation, an agency must therefore consider whether a candidate for regulation is this type of risk.

The comparative approach is appropriate for the two main regulatory hurdles applicable to biotechnology-derived food products, the oversight of the release of the genetically engineered organism during food development and the evaluation of food safety. In neither case are the risks absolutely quantifiable, but they can be compared to the risks of non–genetically engineered food organisms or products in the same situation. Thus, the risk of release of genetically engineered domesticated plants and animals can be compared to the risk associated with the parental or other comparable strain. Genetically engineered food products can be evaluated by comparison to their unmodified counterparts, and if appropriate, to any food additives that might accomplish the same function as the genetic modification. Food, despite its inherent risks, is of course essential, and society accepts many of the environmental risks from its production because of the desirability of traditional agricultural practices. Bioengineered food is certainly a functional substitute for traditional food, and since people's eating habits are unlikely to change dramatically, introduction of the genetically engineered food should not substantially affect total production and consumption of the product, unless there is an exceptional difference in price.

Policy in Transition

Another reason for adopting a comparative approach to regulation of biotechnology-derived food products is that biotechnology represents a small man-made risk superimposed on a background of naturally occurring toxins and carcinogens in food. To eliminate small increments of risk above a large natural baseline is inefficient and costly. The comparative approach also has a sound scientific basis because the view that genetic engineering in the production of food is an extension of long-established conventional breeding techniques underlies the concept that when analyzing risk, genetically engineered products should be readily comparable to their traditional counterparts.

The regulatory framework for bioengineered food is in transition from an approach that, by focusing on the process used to produce genetically engineered food, did not always accurately assess the risk of the product. The Bush Administration sought to cure this problem by adopting a policy similar to the comparative regulatory approach discussed above. The federal regulatory agencies are currently implementing this policy. This policy approach has the advantage of removing unjustified oversight of biotechnology, but it may have the disadvantage of underregulating the field, especially because it relies on existing statutory authority not directed at biotechnology. Moreover, this policy approach does not address the non–risk based social and economic concerns that contribute to the public's objections to biotechnology. It may, therefore, fuel the demand for state and local regulation of biotechnology. . . .

Proposal for Federal Regulation

Regulation of the release and consumption of bioengineered food products requires a credible, straightforward regulatory framework that addresses all genetically engineered organisms and products and gives the states and the public confidence that the products of biotechnology are appropriately regulated. That new statutory authority could extend coverage to all genetically engineered organisms, and it could ease regulatory burdens by coordinating the review process through an applications management board. New statutory authority should be based on the premise that not all products of genetic engineering require oversight. The new statutory scheme should not impose additional regulatory burdens based solely on the process of manufacture or on possible unknown risks; rather, regulation should be commensurate with the risks of the product. Thus, new statutory authority governing the release of genetically engineered organisms could mandate that agencies implement a comparative approach to regulation rather than an undefined standard where permits would be denied only when there is a risk to health or environment. Based on present and future experience, the agencies could develop categories of genetic changes in particular types of organisms that would require only notification, but no review. This would help avoid unnecessary over-

189

sight but provide the authority to regulate when needed.

Recognizing the complexity of the biotechnology industry and its regulation, the extensive technical resources available to the federal review process, and the strong national stake in biotechnology, new statutory authority should also preempt state prohibition of release of genetically engineered organisms allowed by the federal government. It should, however, permit state and local regulation of biotechnology. For example, it should allow states to designate certain areas such as nature preserves as inappropriate for such activities. By giving such regulatory authority, there is the risk that a state could effectively prohibit genetically engineered organisms, perhaps by requiring excessive containment of biotechnology activities. This, however, is somewhat consonant with other federal regulatory statutes that allow states to regulate more stringently than federal statutes require. Because biotechnology is a rapidly changing area, any new statutory authority should contain a sunset provision to allow change that will accommodate new advances and understanding of risks.

New Foods Similar to Traditional Foods

Since the food products of biotechnology are comparable to traditional foods, they could be regulated under existing statutes, but new regulations addressing generic concerns related to biotechnology would be helpful to developers and regulators. This proposal is consistent with the recommendation of the Administrative Conference of the United States that agencies adopt rules that address recurring regulatory issues concerning biotechnology. Although the food additive and GRAS [generally recognized as safe] categories may be appropriate for evaluating bioengineered chemicals and enzymes, transgenic food crops should be evaluated on a comparative basis without reference to whether the gene/product might be considered an added substance or food additive. The provisions could still be self-actuating in the event that manufacturers are confident that their foods meet the specifications, but notification should be mandatory during this period when bioengineered foods are still being introduced. The agencies should continue the practice of giving advisory opinions when manufacturers request advice on the status of their product. Federal regulation of biotechnology-derived food products would not preempt state law, since state law protecting health and safety is generally given considerable deference where there are no explicit preemption provisions.

In conclusion, a more comprehensive and straightforward federal statutory and regulatory scheme, addressing both the environmental release of bioengineered food organisms and the safety of the products, could foster both the development of biotechnology and protection of human health and the environment. It should not, however, thwart state or local laws that have long been accepted in the area of food regulation. The only remedies, then, to unduly restrictive regulation at the state level lie in the credibility of the federal scheme and, more importantly, in the public perception of the technology itself. That perception will be greatly improved when biotechnology advances to the point that products are available to fulfill its promise of improvement of human health and the environment.

Editor's note:

This article was not written in the author's capacity as a government official. Any opinions, findings, conclusions, or recommendations expressed in this article are those of the author, and do not necessarily represent the views of the Office of Inspector General, the National Science Foundation, or the National Science Board.

"To believe that the industry can regulate itself and the government can ensure safe, humane, and environmentally neutral . . . applications of biotechnology is simply wishful thinking."

Federal Regulation of Genetic Engineering Is Ineffective

Michael W. Fox

Michael W. Fox is vice president of the Humane Society of the United States and has spearheaded the movement to foster the ethical treatment of animals. He is the author of many books, including *The Soul of the Wolf*. In the following viewpoint, Fox outlines government efforts at regulating genetic engineering. He concludes that these efforts have failed, and that neither the government nor the genetic engineering industry can be trusted to protect consumers and the environment from abuses of this new technology.

As you read, consider the following questions:

1. What conclusions did the Government Accounting Office make in 1986 concerning the U.S. Department of Agriculture's regulation of biotechnology, according to Fox?
2. What problems does the author see in the Environmental Protection Agency's regulation of genetically engineered microorganisms?
3. Why is public involvement concerning the regulation of genetic engineering important, in Fox's opinion?

Francis Bacon cautioned that if we would control nature, we must first obey her. Today we understand obedience as respect for nature's "laws." The most important aspect of this respect is being responsible for the environmental and ecological consequences of our actions. This has now become written into our law, which, for better or worse, is administered by the Environmental Protection Agency (EPA).

This agency recently expressed confidence that the safety and environmental impact of genetically engineered organisms can be adequately determined and regulated. Such confidence is undermined by the EPA's historically documented inability (along with the U.S. Department of Agriculture) to protect the environment and the public's health from the wholesale misapplication of petrochemical-based agripoisons (pesticides, herbicides, and fungicides). Although these old agrichemicals now contaminate our food, water, and body tissues, at least they did not have the capacity to multiply; the new bacterial pesticides do.

Clearly genetic engineering is as much a Pandora's box as it is a cornucopia of wonderful possibilities. Like any other product of human ingenuity, it has great potential risks as well as benefits to society. And for those who invest in this new industry—since genetically engineered plants and microorganisms can be patented—fortunes can be made. But without congressional and state oversight and international coordination to minimize risks to the environment and to the very fabric of life itself, we could be on the threshold not of some biological utopia, but of our own nemesis.

The Watchdogs

The blossoming biotechnology industry is now regulated by already existing governmental regulatory agencies that have been jockeying for jurisdiction over various sections of the industry (see Table 1). But how effective can these agencies be?

The present governmental regulatory framework for the U.S. biotechnology industry is based on the Federal Policy on Biotechnology. Dr. James W. Glosser, chief administrator of the U.S. Department of Agriculture's (USDA) Animal and Plant Health Inspection Service (APHIS), summarized it as follows:

> The Federal Policy on Biotechnology was established December 31, 1984, and published in final form on June 26, 1986, by the U.S. Office of Science and Technology Policy (OSTP) as the "Coordinated Framework for Regulation of Biotechnology." The OSTP concluded that products of recombinant DNA technology will not differ fundamentally from unmodified organisms or from conventional products. Therefore, the existing laws and programs are adequate for regulating organisms and products developed by this process.

Table 1
Federal Agencies Responsible for the
Approval of Biotechnology Products

Biotechnology Product	Responsible Agency
Pesticide microorganisms released in the environment	EPA APHIS
Other uses of microorganisms: Intergeneric combination	EPA APHIS
Foods/food additives	FDA FSIS
Human drugs, medical devices, and biologics	FDA
Animal drugs	FDA
Animal biologics	APHIS
Other contained uses	APHIS FSIS FDA

APHIS: Animal and Plant Health Inspection Service
EPA: Environmental Protection Agency
FDA: Food and Drug Administration
FSIS: Food Safety and Inspection Service

Source: Michael W. Fox, *Superpigs and Wondercorn*, 1992.

Many would not agree with this conclusion, which is the cornerstone of the U.S. government's oversight and regulation of the biotechnology industry. Dr. Glosser went on to state:

> The Coordinated Framework included an index of laws applicable to biotechnological products in the various stages of research, development, marketing, shipment, use, and disposal. The framework also included policy statements from the three federal agencies that share the major responsibilities for regulating products of recombinant DNA technology—the Food and Drug Administration, the Environmental Protection Agency, and the USDA. . . .

> A key element in our structure was creation of a special staff to coordinate biotechnology regulatory activities for the Department. The staff serves as a liaison with other government agencies, industry, and the general public on USDA's regulation of biotechnology. The staff actively participates in the United States effort to promote international consistency on biotechnology regulations. Such international consistency should prevent unnecessary trade barriers and ease the transfer of American products into international markets.

It is noteworthy that the emphasis here is on promoting inter-national regulatory consistency not to minimize adverse envi-ronmental consequences, but rather to facilitate the global ex-pansion of U.S.-based biotechnologies.

Failure of the USDA

After a year's study of biotechnology regulation at the USDA, the Government Accounting Office reached three main conclu-sions in its report released on 3 April 1986:

1. The main biotechnology committee at the USDA—the Agri-culture Recombinant DNA Research Committee—has almost no authority, meets infrequently, has no budget, and its meeting records show confusion about what the committee should do.

2. The department has no clear policy about who should review biotechnology proposals and what rules should be applied. The report stated, "Different agencies in USDA have been jockeying for regulatory control, and USDA officials have expressed uncer-tainty as to which agency is responsible for different activities."

3. The USDA has done little to communicate to Congress and the public the benefits and risks of biotechnology.

The mandate of this chronically understaffed and underfunded federal agency (the USDA) to protect animal and plant health clearly omits reference to environmental protection, which is the domain of the EPA. Some would argue that the EPA should regulate all environmental releases of biotechnology products, since the USDA was so ineffectual in regulating the pesticide in-dustry that the responsibility was transferred to the EPA.

How can the USDA be expected to effectively regulate the re-lease of engineered plants, microorganisms, and animals (includ-ing insects and other plant pests) as well as animal drugs and live-virus vaccines developed through genetic engineering, con-sidering its past failure with pesticides? But it is doubtful that any federal agency can effectively regulate this new industry.

FDA Failures

The Food and Drug Administration (FDA) regulates new drugs and vaccines developed from biotechnology. But the existing reg-ulations and test protocols for all new pharmaceuticals are inad-equate and inappropriate, according to a recent Government Accounting Office report. For example, growth hormone manu-factured from genetically engineered bacteria passed all the rou-tine assays and toxicology tests but had some unanticipated and significant clinical side effects, which were subsequently found to have been caused by unidentified contaminants. The public has not forgotten the thalidomide tragedy (when animal tests run before its approval for human use proved, retrospectively, to be inappropriate and invalid) nor the recent tragedy caused by genetically engineered L-tryptophan that was thought to be safe.

The FDA is also in charge of monitoring our food for contamination with agrichemicals, drugs, and bacteria, along with the USDA's meat inspectorate. Data indicating high levels of agrichemical residues in imported and domestic foods cast doubt on the FDA's ability to regulate the agribusiness food industry effectively if and when biotechnology, as well as food irradiation, become integral components. . . .

The Failures of Other Departments

The EPA has been regulating the release of genetically engineered microorganisms intended for use as pesticides under the Federal Insecticide, Fungicide and Rodenticide Act (FIFRA) and for other purposes under the Toxic Substances Control Act (TSCA). Both these acts were intended to regulate chemical pesticides, not living, genetically engineered life forms. Many analysts consider them wholly inadequate for this purpose, and I wholly concur. According to the OTA [Office of Technology Assessment], the EPA now works with APHIS in reviewing industry and university applications under FIFRA to release microorganism pesticides. As of March 1991, EPA had approved ten applications for small-scale testing of genetically engineered microbial pesticides under FIFRA. In addition, two applications had been withdrawn, and another review had been suspended.

The Recombinant DNA Advisory Committee of the National Institutes of Health has established guidelines for research. But some of the committee members believe that the existing guidelines should be abolished because there have been no accidents involving genetically engineered bacteria and the public health since the committee's inception in 1974. What is more disturbing is that this same committee ignored its own guidelines by permitting University of California agriculturalists to release genetically engineered bacteria into the environment. . . .

The Public's View of Biotechnology

The hallmark of a technocratic society is the appropriation of the public right of involvement and ultimately of democratic self-determination. And when children are raised to obey the voice of authoritarian rationalism without question despite what they feel and know intuitively, the viability of technocracy is assured—at least for a few generations, until either the human spirit rises up in revolt or the system disintegrates under the combined weight of unmanageable complexity and bureaucratic-managerial entropy. Hence, public involvement in the policy decision making of the biotechnology industry is important at all levels, both nationally and internationally.

Because the general public is being led by the biotech industry to believe that biotechnology is a panacea for many of life's problems, it is difficult to take the position of a critic. Such a position seems both antiprogress and antisocial. To express concern for animals' welfare or nature's creation is seen as relegating human interests to the backseat. It is easier for the industry to gain public support by arousing hope and faith in science than for its critics to appeal to reason and evoke public fear and doubt. No one wants to bear the burdens of fear and concern for the adverse consequences of biotechnology. And to believe that the industry can regulate itself and the government can ensure safe, humane, and environmentally neutral (if not enhancing) applications of biotechnology is simply wishful thinking.

"Genetic manipulation should be monitored by a governmental board or agency composed of members from all aspects of human life."

A Unified System of Federal Regulation Is Needed

Geoffrey Baskerville

By manipulating genes, scientists can treat and possibly cure many genetic disorders, Geoffrey Baskerville explains in the following viewpoint. Unfortunately, the same technology that offers such promise also poses threats to society, the author writes. Baskerville proposes that the U.S. Congress establish a committee of scientists, ethicists, and people from the general public to make recommendations concerning the safe use of genetic engineering. Baskerville is an attorney in Philadelphia, Pennsylvania.

As you read, consider the following questions:

1. What tension is created by the idea that science is devoted to the betterment of human beings, according to Baskerville?
2. What are the guidelines for genetic engineering established by the National Institutes of Health, according to the author?
3. How would Baskerville change the current regulatory system?

Excerpted from "Human Gene Therapy: Application, Ethics, and Regulation" by Geoffrey Baskerville, *Dickinson Law Review*, vol. 96, no. 4 (Summer 1992), pp. 733-763. Reprinted with permission of the *Dickinson Law Review*.

On September 14, 1990, Drs. R. Michael Blaese, W. French Anderson and Kenneth W. Culver opened the door on a new age of medical treatment when they injected a four-year-old girl who suffers from adenosine deaminase deficiency (ADA), a rare genetic disorder, with blood cells containing genetic material designed to correct her genetic disease. . . . It is the hope of the doctors that the new genetic material injected into the girl's blood will begin to produce copies of the ADA gene and, in so doing, create a functioning immune system. This experiment marks the first time that doctors have attempted to combat a genetic disease in humans through the use of gene therapy. As scientists learn more about the human genome and the genetic basis of disease, the number of applications for this type of therapy will grow. Researchers already have announced the discovery of the genetic origins of diseases such as sickle cell anemia, cystic fibrosis, and osteoarthritis. Recent studies also indicate that diabetes, Alzheimer's disease, and all types of tumors have a genetic basis.

In addition, on January 29, 1991, Dr. Steven A. Rosenberg performed gene therapy on two patients suffering from metastatic melanoma, a lethal type of skin cancer. The process used in the melanoma patients was similar to that used in the patient with ADA except that the gene injected in the cancer study is designed to produce a tumor-destroying enzyme. Recently, Dr. Rosenberg began a second experimental procedure to combat malignant melanoma. In this new procedure, Dr. Rosenberg removes cells from the patient's tumors, adds genes from a naturally occurring anti-tumor toxin, and then injects the cells back into the patient. After three weeks, white blood cells, or lymphocytes, are removed from the area of the tumor. These modified white blood cells, believed to be the body's most potent tumor fighting agent, are duplicated in the laboratory and then injected into the patient.

Gene therapy is just one of the possible applications of genetic engineering or biotechnology, the terms by which the human alteration of deoxyribonucleic acid (DNA) have come to be known. Other applications of this technology range from the production of certain types of drugs to the creation of plants and animals with specific characteristics. All of these applications offer great promise and benefits for mankind; yet, many observers find this new technology ethically troubling, especially when the technology is applied to human beings. One need not imagine a Brave New World or a Frankenstein to see the potential for abuse of genetic engineering. . . .

Science and Regulation

There can be no doubt that, as Daniel Callahan states, "[s]cience is now a public and social enterprise not only because much of its financial support comes from the public but just as importantly because the implications and impact of science are public and social." This is especially true for genetic engineering. Therefore, it is argued that because of the public and social implications and impact of science, particularly genetic manipulation, government regulation of science is necessary.

"The good news, however, is that thanks to Dr. Prebish we can implant in you the genes of someone I won't yearn to fire."

Reprinted by permission of Andrew Toos.

Many scientists, however, feel that the best science is that which is conducted without social constraint. To this end "[m]any scientists claim . . . [that] they have a right of free inquiry—a right to research—that governmental constraints on scientists' choice of research topics and methods violates," John A. Robertson writes. Thus, the idea that science is an enterprise devoted to the betterment of human beings creates a tension between the good of society and the freedom to research. There must, of course, be limits imposed on any freedom when the exercise of that freedom will harm the health, safety, and welfare of others. Therefore, the state may regulate research because society may be harmed by either the research process itself or the product of the research. In addition, it must be realized that the interests of the public and the interests of scientists will not always be the same. Therefore, we must concede as Callahan does that "[s]cientists cannot fully regulate themselves in the public interest because they are not necessarily representative of the public." Despite the initial success of self-regulation by scientists in the field of genetic engineering, the need for outside regulation is apparent due to the severe consequences that could arise if the process was abused.

Even with a system of governmental regulation, some scientists will attempt to avoid the established protocol and conduct research as they please. This is clearly shown by the incident in 1980 involving University of California–Los Angeles [UCLA] researcher Martin Cline. Dr. Cline, as Ira H. Carmen states, "will go down in medical history as the first person to attempt to transplant cloned genes into human patients, his particular purpose being to effect a cure for the dreaded globin disease beta thalassemia." Dr. Cline performed the experiments in Italy and Israel, but failed to receive approval from either government or from the necessary review boards at UCLA.

When the violations were reported, the National Institutes of Health (NIH) conducted a review of the matter. The NIH concluded that Dr. Cline was in violation of the established procedure and removed federal funding from Dr. Cline's projects. In a more recent incident, one of America's leading molecular biologists, Dr. David Baltimore, has been implicated in the cover-up and use of falsified data concerning the effect of transferred genes on the human immune system. Dr. Baltimore did not conduct the research in question, but he was the co-author of the paper in which it appeared, and he initially defended its contents. Dr. Cline's failure to gain proper approval for his studies and Dr. Baltimore's initial reaction to the criticism of his paper are examples of incidents in which a scientist has been too involved in his/her research to be able to view the research objectively. Therefore, in order to benefit society, the application of genetic manipulation should be monitored by a governmental board or agency composed of members from all aspects of human life.

The Present Regulatory System

The present system of regulation for rDNA technology began in 1974 when the NIH established the Recombinant DNA Advisory Committee (RAC). The RAC is composed of twenty-five members drawn from the fields of law, molecular biology, medicine, ethics, and the general public; its initial task was to develop procedural standards for NIH-funded rDNA experiments. These standards, the Federal Guidelines for Research Involving Recombinant DNA Molecules (NIH Guidelines), were published in 1976 and have since been revised. The central focus of the NIH Guidelines is to prevent environmental contamination by the rDNA-altered organism used in the experiment. In general, the NIH Guidelines establish levels of physical and biological containment necessary for the safe handling of genetically altered organisms. In cases where serious risk is presented, scientists must submit detailed data, including toxicity studies, to the RAC on the host/vector system that they intend to use. The RAC makes its recommendation and then passes the application on to the director of NIH for final approval. Finally, the NIH Guidelines require an institution conducting even low-risk experiments to establish an Institutional Biosafety Committee (IBC) to ensure that the procedures established in the NIH Guidelines are followed.

Regulation of Human Gene Therapy

In 1984, the RAC realized that the initial guidelines had not fully taken into account the possibility of human applications of genetic manipulation. Therefore, it established the Human Gene Therapy Subcommittee to explore the specific problems of human gene therapy. The Subcommittee proposed and the RAC approved a document titled "Points to Consider in the Design and Submission of Human Somatic-Cell Gene Therapy Protocols," which, according to Judith Areen and Patricia King:

> formally requests . . . [a researcher to provide] information on a variety of topics that the Human Gene Therapy Subcommittee considers relevant to the task of deciding whether to recommend approval of a specific protocol for human gene therapy; implicitly, however, it imposes requirements on researchers seeking approval that go beyond the RAC Guidelines.

In addition, human subjects in rDNA research experiments are covered by the general federal regulations concerning the use of human beings in all experiments. Under these regulations, each research facility is required to establish an Institutional Review Board (IRB) to review protocols using human subjects. Approval for such research is given only if the risk to the human participants is minimal or in proportion to the potential benefit from the protocol. The human subjects or their legal representative must also give informed consent to the process. Therefore, as Areen and King state, "[a]ny protocol for human gene therapy conducted at an institution receiving NIH research funds will need the approval of at least four different regulatory bodies before going forward: the local IBC, the local IRB, the RAC and the FDA [Food and Drug Administration]."

This impressive array of regulatory authority is deemed necessary to protect the public at large and the individual research
subjects from any possible ethical or ecological harm resulting
from the use of gene therapy. Yet, in spite of all its internal
checks and balances, the system has some serious problems. To
begin with, the NIH Guidelines apply only to institutions that
receive funding from the NIH. In addition, the legal basis for
the NIH's power to require compliance with the NIH Guidelines even for NIH-funded institutions is somewhat unclear. . . .

Potential for Conflict

Another problem with the present regulatory system is that
the NIH is not the only government agency involved in the regulation of genetic engineering. Other agencies with regulatory
concerns relating to rDNA are the Food and Drug Administration (FDA), the Environmental Protection Agency (EPA), the
Department of Agriculture (USDA), the National Science Foundation (NSF), and the Occupational Safety and Health Administration (OSHA). While many of these agencies have incorporated
the NIH Guidelines into their individual regulatory schemes, the
potential for conflicting decisions is still present. The following
analysis will concern only the interaction between the FDA and
the NIH in the regulation of gene therapy.

Beginning in 1979, the FDA asserted that its regulatory powers
extended to all new drugs and methods of producing new drugs.
The FDA interpreted this to mean that even when a drug made
by conventional means had been approved for use, the same
drug when made by another company or by different means
would be a "new drug." Therefore, generic drugs or drugs produced by rDNA processes were to be included under the FDA's
pre-marketing review process. The United States Supreme Court
ultimately upheld the FDA's extension of power. In addition,
the FDA claims authority to regulate human gene therapy
through its power to regulate all clinical pharmaceutical testing.
This proposition presents two problems for the scientific community and the public at large. First, the FDA's enabling statute
gives the agency power to regulate drugs and products used in
clinical testing when they move in interstate commerce. Therefore, it is possible to question the FDA's authority to regulate
gene therapy if the rDNA product given to the patient has not
moved in interstate commerce. Second, some commentators
question the FDA's ability to require compliance with the NIH
Guidelines.

Under the basic principles of administrative law, an agency cannot promulgate substantive or interpretive regulations unless the
regulations advance some provision of the agency's enabling

statute. If FDA were to incorporate the NIH Guidelines, the commentators suggest that the only logical way to do so would be under the Good Manufacturing Practices regulations (GMP) of the Food, Drug and Cosmetic Act. The GMP regulations carry the force of law and cover a wide range of practices, including those conducted in laboratories as well as those conducted in packaging and labeling. The problem with incorporating the NIH Guidelines into the GMP regulations is that the objectives of the two systems are quite different. The GMP regulations are designed to carry out the overall goal of the Food, Drug and Cosmetic Act—purity of food and drugs to ensure the safety of the people who use them—while the NIH Guidelines were established to "specify practices for conducting basic research and have nothing to do with assuring the identity, strength, quality and purity of the products of the technology." Therefore, it appears that FDA would be going outside its administrative mandate in attempting to require compliance with the NIH Guidelines.

Attempts to Amend the Present System

The different purposes of the NIH and the FDA highlight the potential for conflict in regulating human gene therapy due to the overlapping jurisdiction of the agencies involved. These jurisdictional conflicts will also lead to inconsistent regulations being issued by the different agencies. Therefore, it is clear that some method of resolving the present difficulties in regulating genetic engineering must be found. The government took steps in this direction when the White House Office of Science and Technology created the Biotechnology and Science Coordinating Committee (BSCC) in 1985. The BSCC was designed to provide all the agencies regulating genetic engineering with a forum in which they could discuss the scientific and regulatory problems created by the new technology. In this way, general scientific understanding could be reached and all agencies would be able to regulate even-handedly. The problem with the BSCC is that it was given no power to resolve interagency disputes or to enforce its recommendations. Therefore, the BSCC lacks the authority to be a centralized regulatory overseer for the developing applications of biotechnology.

Another possible solution to the regulatory problems created by genetic engineering is to elevate the status of the RAC and Human Gene Therapy Subcommittee to a position in which they could oversee all aspects of genetic engineering. This also creates some problems, specifically for the medical applications of genetic engineering, because the RAC and its subcommittee do not have a mandate to address the ethical issues presented by either germ-line gene therapy or enhancement genetic engineering. Therefore, Adrienne Naumann and other commentators have proposed the establishment of a new "super agency" or commission created in the mirror image of the RAC, but with rule-making authority and the power to settle interagency disputes.

Congress also became directly involved in the monitoring of biotechnology in 1985 when it enacted legislation allowing for the creation of a Bioethics Board (BEB). The BEB was modeled on the Office of Technology Assessment and was comprised of twelve members (six from each House of Congress—an equal number from each party). The function of the BEB was to provide Congress with a continuing report on the ethical issues arising in the new field of biotechnology. To aid the BEB in this task, the board was to appoint a fourteen member Biomedical Ethics Advisory Committee (BEAC) composed of four scientists, three doctors, five ethicists, and two members of the public with an interest in biomedical ethics. The promising potential of the BEB and BEAC all came to naught, however, when the BEAC could not agree on a replacement for one of the original members who had died before the first meeting. Finally, Congress withdrew funding for both the BEB and BEAC in 1989. . . .

A Unified System Is Needed

The collapse of the BEB and BEAC in 1989, combined with the inability of the BSCC to resolve interagency disputes, creates a serious obstacle to any unified policy of regulation for genetic engineering. Without a unified and consistent policy of regulation, the progress of genetic engineering, especially human gene therapy, may be slowed by overregulation. It is essential for the development of human gene therapy, therefore, that the government overcome the obstacles that it has created and devise a unified system of regulation. This could best be accomplished by Congress through the refunding of the BEB and BEAC. The BEB's function could be expanded to include a review of the present regulatory system and the development of options for streamlining the regulatory process. In this way, the BEB could address both the ethical and regulatory problems presented by the continued application of human gene therapy. Such a committee would be endowed with the best elements of the present RAC and the ability to resolve interagency disputes that is lacking from the BSCC.

Present members of the RAC and members of the agencies currently regulating genetic manipulation would be ideal candidates for positions on the BEAC. This would ensure that the decisions reached by the BEAC would be the work of scientists, ethicists, and the general public. In addition, the overall recommendations of the BEB would carry the weight of a congressional committee report. Congress could then enact the necessary legislation to implement the BEB's recommendations. Once the new system of regulation was in place, the BEB would continue to report to Congress on the progress of the new system and the evolving concerns presented by the development of human gene therapy. The end result of such a process would be a unified system of regulation for gene therapy that would avoid the problems presented by overlapping agency jurisdictions and a permanent Congressional committee to oversee the regulatory process and resolve any regulatory or ethical problems that arise in the development of human gene therapy. Even if the BEB recommended the continuation of the present system of regulation, the BEB could still function as the information gathering and reporting mechanism that Congress originally intended it to be. A functioning BEB would ensure that Congress is kept well-informed about the difficult issues that will arise as human gene therapy develops. Therefore, it is clear that Congress should redesign and refund the BEB and BEAC, for it is through the operation of these two boards that the government will best be able to respond to the challenges of biotechnology and human gene therapy.

"The wisest course would be a self-imposed but strictly observed moratorium by insurers on any genetic screening of applicants."

Genetic Testing by Insurers Should Be Self-Regulated

Alexander Morgan Capron

Alexander Morgan Capron is Henry W. Bruce University Professor of Law and Medicine at the University of Southern California in Los Angeles. In the following viewpoint, Capron writes that insurance companies are now screening, or thinking of screening, applicants for genetic illnesses. The author believes this trend could lead to discrimination against those who test positive for genetic disease. He concludes that insurers should observe a self-imposed moratorium on the genetic screening of applicants and that research on such testing should adhere to ethical guidelines.

As you read, consider the following questions:

1. What possible breaches of law and ethics were committed by the Transamerica Occidental Life Insurance Company, according to Capron?
2. Explain the author's view that insurance companies and the public both need to be protected from the consequences of genetic testing.
3. What are some of the laws states have passed concerning genetic illness and discrimination, according to Capron?

Alexander Morgan Capron, "Hedging Their Bets," *Hastings Center Report*, May/June 1993. Copyright ©1993 The Hastings Center. Reproduced by permission.

Predictive medicine—monitoring and even treating potential illness in people who are asymptomatic—holds the promise of transforming health care. Armed with foreknowledge of individuals' particular susceptibilities, physicians will be able not only to intervene before disease takes its toll but also to tailor the advice they give patients about diet and life-style, focusing attention on the salient measures that now get lost in the deluge of generalized guidance and admonitions.

Yet this new field holds peril as well as promise since findings about disease susceptibility may interest people other than the ones being tested. Indeed, the possible interest of insurance companies in such data has been a point of concern for those who see a double-edged sword in the techniques being developed by the genome mappers and others on the frontiers of molecular medicine.

Several states have responded by enacting legislation to restrict the uses that insurers can make of "genetic information." In opposing such laws, the insurance industry has claimed that they are unnecessary because insurers have no plans to make use of genetic tests now. The concern that genetic results will render some people unable to obtain health insurance reflects underlying problems in health care coverage and would best be addressed as part of current efforts to reform access to health care generally.

Using Applicants as Experimental Subjects

Nonetheless, one major life insurance company's wholesale use of an experimental screening test not only suggests that at least some insurers are interested in predictive medicine but also raises questions about how reliability data for such screening devices will be gathered. On 14 January 1993 the California Department of Insurance opened an investigation of Transamerica Occidental Life Insurance Company's use of a new blood test for cancer antibodies to screen over 50,000 applicants over the previous two years. The department is looking into allegations, denied by the company, that the test was used to rate or deny applicants for insurance policies.

The willingness of Transamerica to use a test characterized by a scientist at the National Cancer Institute as unreliable and inappropriate for screening throws doubt onto the insurance industry's insistence that genetic tests are not being used since they are obviously still just experimental. The company—like any thinking of adopting a new screening tool—understandably wanted to be able to track whether the test provided reliable information for underwriting purposes. But the situation is complicated for Transamerica because it owns 30 percent of the laboratory that developed the test and hopes, if the test pans out, to market it to other insurers. Thus it appears to have used its applicants as experimental subjects.

Furthermore, the company admits that applicants were not told about the test but merely signed forms consenting to blood tests for "tumors." Again, this is problematic for Transamerica because California is one of the few states with a statute that establishes standards for human experimentation. That statute requires explicit, written informed consent after subjects have been given a copy of the "experimental subject's bill of rights."

Testing Must Be Strictly Controlled

Ultimately, the decision to use genetic testing will be hampered by the lack of understanding by most physicians and health care professionals. Companies marketing these tests will nevertheless target physicians and consumers and thereby speed the adoption of testing. We may find ourselves in a situation where health insurance companies will view genetic screening as a means of cost avoidance. . . . The testing systems in use must be improved, and validity and reliability guaranteed. In order to insure personal freedom, strict control of genetic testing by insurers and employers is a must.

Michael L. Begleiter, *Midwest Medical Ethics*, Summer 1992.

It would be surprising were insurers totally uninterested in the torrent of predictive tests that the life sciences are starting to unleash. Some tests rely on older methods, such as antibody detection, but the most eagerly awaited are those that look for genetic variation at a molecular level rather than searching for cellular products. While many genetic conditions are so rare that population screening may never be practical (unless technical advances reduce the cost to pennies per test), some diseases are more common, and even for rare ones a genetic test may be justified economically when an applicant wants a very large policy.

Recognizing that premature death or severe illness could have a catastrophic effect on their families, people are willing to pay a small amount to cushion the financial consequences of misfortune. So long as an insurance pool is large enough and participants have not joined because of atypical peril, their collective risk should correspond to the actuarial data on which their premiums were established, and the pool should have enough resources to fund the losses of its members.

Accordingly, insurers try not only to enroll large numbers but also to avoid situations in which those purchasing insurance have greater knowledge about their probability of experiencing a loss than the insurer has. Since predictive medicine involves just such a potential differential in knowledge, insurers have an obvious interest in learning any information of this type that has been provided to policy applicants.

Testing for AIDS

If insurers are unable to take this information into account to set a person's premiums, or perhaps even to decline to write a policy, they fear issuing policies disproportionately to people who know they are more likely than average to experience an insured event—what is known as "adverse selection." In competitive terms, companies particularly dread being the only one that does not screen out such adverse selectors.

The most dramatic recent example of the industry's fears in this regard was provoked by the AIDS epidemic. While people in their twenties and thirties are usually eagerly sought by insurers and are charged low premiums, companies found themselves potentially insuring people who believed themselves to be at risk of being infected by HIV (or who had even already been infected) and whose insurable losses would occur at much higher rates than anticipated for the general population. As one would expect, insurers wanted to screen applicants for HIV. Their desire provoked attempts (successful in several jurisdictions) to bar AIDS testing or at least to insist that tests be performed only with the explicit, informed consent of the persons tested. It also generated a lively debate over the purposes of various types of insurance and the appropriate role of the means used by insurers to differentiate among applicants based on various perceived risks.

The industry claimed that state laws mandating fairness obliged them to treat people with HIV infection differently from other people because failure to do so would mean that healthy people would be charged higher premiums than could be justified based upon their expected losses. Critics replied that the industry already allowed certain factors to override actuarial considerations, and further that society sometimes insists that other considerations, such as gender neutrality, take precedence.

No one disputed, however, that apart from social insurance programs (those that reach an entire population and that typically are funded through taxes or comparable mandatory contributions), insurance plans must be able to take certain predictors of loss into account if they are to be actuarially sound. This need is especially strong when a new illness arises that significantly changes expectations—in terms, for example, of medical costs and death rates.

Such has been the case with AIDS since the early 1980s; such is not the case, however, with most diseases that will be susceptible to the capabilities of predictive medicine. Rather, these conditions (such as genetic illnesses and cancer) are already included in insurers' actuarial tables. The new molecular tests may simply permit the probable victims of such conditions to be identified much earlier and more precisely.

The dilemma is thus clearly posed: if people are going to undergo predictive screening, both to advance knowledge in the field and to find out information of potential value to their own health and even their own life, they will need to be protected against unfair discrimination by insurance companies, while at the same time these companies will need to be protected against adverse selection.

Outlawing Genetic Discrimination

Several states have enacted laws that explicitly limit the use of genetic tests by insurers. But the actual effect of such laws on insurance may be much more modest than the statutes suggest on their face. Since the 1970s, laws against discrimination based on sickle-cell anemia and several other genetic diseases or carrier states have been on the statute books. In 1989 Arizona became the first state to enact protection based on "genetic conditions" generally, but it merely brought such conditions within the prohibition of unfair discrimination, while permitting insurers to consider genetic risks that substantially affect actuarial predictions.

Florida took the stronger step of requiring informed consent for DNA analyses, except in criminal investigations. Results may not be disclosed without the consent of the person tested, but insurers are apparently free to use them to determine eligibility and rates, provided that the tester reveals this to the person tested. Should insurance be denied, the test "must be repeated to verify the accuracy."

Probably the strongest statute is the one that became effective 1 July 1992 in Wisconsin. It provides a preview of what are certain to be an increasing number of efforts to limit "genetic testing," defined there as a test using deoxyribonucleic acid extracted from an individual's cells in order to determine the presence of a genetic disease or disorder or the individual's predisposition for a particular genetic disease or disorder.

Under the statute, insurers (as well as employers that self-insure) are not only prohibited from requiring genetic tests but also from requesting information about previous tests. Furthermore, coverage may not be conditioned upon having a genetic test, nor may rates be determined upon test results. None of these restrictions apply, however, to life insurers, which are mandated only to behave reasonably when setting rates based on genetic data.

This willingness to allow life insurers to rely on predictive information while excluding it for health coverage is not surprising. Not only is health insurance regarded as more essential but because most of it is written in group policies, the whole notion of underwriting is under heavy attack in many quarters.

For the moment, the wisest course would be a self-imposed but strictly observed moratorium by insurers on any genetic screening of applicants, coupled with equal restraint on the part of state legislatures in enacting formal prohibitions regarding insurers' use of the new means of predictive medicine. Furthermore, research on the usefulness of the new tests ought to be carried out by scientists operating independently of the insurance industry, openly and in accord with ethical guidelines for research, including the subjects' informed, voluntary consent.

"I think it would be a serious mistake to adopt policies aimed at preventing the development of a technology capable of making major modifications in the human genome."

Genetic Engineering of Humans Should Not Be Regulated

Stephen P. Stich

Stephen P. Stich is a professor of philosophy and cognitive science at Rutgers, the State University of New Jersey, in New Brunswick. In the following viewpoint, he writes that genetic engineering offers the potential to greatly improve the human species. Rather than regulation, he advocates the formation of an independent body of scientists, ethicists, religious leaders, and others to study developments in genetic engineering. He also calls for increased public education about the benefits of the new technology.

As you read, consider the following questions:

1. What problems does Stich foresee as genetic engineering advances and becomes more commonly used?
2. What two arguments have led people to oppose manipulating the human genome, according to the author? How does he refute these arguments?
3. What example does Stich give to support his contention that the results of natural selection can be improved upon?

Stephen P. Stich, "The Genetic Adventure," in *Values & Public Policy*, edited by Claudia Mills. Orlando, FL: Harcourt Brace, 1992. This article originally appeared in the Spring 1983 issue of *QQ: Report from the Institute of Philosophy and Public Policy* and is reprinted here with the author's permission.

Humankind is embarked on an extraordinary adventure, an adventure promising rewards that could barely have been imagined as recently as a generation ago. But human genetic engineering poses moral and social dilemmas every bit as daunting as the rewards are enticing.

It seems clear that in the decades ahead research yielding knowledge relevant to human genetic engineering will continue and accelerate. What is more, I think the acceleration of research in this area is to be welcomed. The medical, industrial, and agricultural applications of genetic engineering research will transform our society in ways even more profound than the computer revolution now well under way. I am enough of an optimist to believe most of these changes will be for the good. However, as we learn more about the mechanisms of human genetics, it is also inevitable that we will start learning how to manipulate the human genome to suit our tastes, or what we perceive to be our needs, in domains far removed from those that traditionally have been the concern of medicine.

The Improved Human Being

At first this ability will be restricted to characteristics under the control of a single gene or a small number of genes. But as our knowledge progresses, we will learn more and more about how to manipulate those characteristics of beings—both physical and mental—under the control of many separate genes. In our current state of knowledge we simply do not know the extent to which aspects of intelligence and personality are under genetic control and thus susceptible to genetic manipulation. But as I read the evidence we now have, there is every reason to think that a substantial component of our mental and moral lives is influenced by our genetic endowment. It would be remarkable indeed if we did not all come genetically equipped with mental strengths and weaknesses just as we come equipped with innate physical strengths and limitations.

As our ability to manipulate the genetic composition of our own offspring grows in sophistication in the decades ahead, the social pressures to use this new technology will become intense. Since the early 1980s we have seen an explosion of interest in home microcomputers; many of the people who buy these wonderful, expensive machines do so in the hope they will give their children a competitive edge in a technologically competitive world. Closer to the fringes of our society we have seen that some women are prepared to have themselves impregnated with the sperm of a Nobel Prize winner in the hope of bearing an intellectually gifted child. Both of these phenomena underscore the fact that the desire to help one's children excel is a powerful and widespread motivational force in our society. When, via genetic engineering, we learn how to increase intelligence, memory, longevity, or other traits conveying a competitive advantage, it is clear that there will be no shortage of customers ready to take their place in line. Moreover, those who are unwilling or unable to take advantage of the new technology may find that their offspring have been condemned to a sort of second-class citizenship in a world where what had been within the range of the normal gradually slips into the domain of the subnormal.

Expanding Genetics Instruction

The public needs to become educated about issues of genetic screening and testing. Activities aimed at achieving this goal have been stated. The years since the early 1980s have seen an expansion of genetics instruction beginning at the elementary school level and continuing through high school. At least one federally funded project is training teachers in human genetics. The plan is for each of these individuals to serve as lead teachers in their school districts to help their colleagues develop the knowledge necessary to teach about this rapidly changing area.

Michael L. Begleiter, *Midwest Medical Ethics*, Summer 1992.

Obviously if history unfolds more or less along the lines I have been predicting, plenty of social problems will be generated. Ensuring equitable access to the new technology and protecting the rights of parents and children who have chosen not to utilize the technology are two that come quickly to mind. These issues, however, are variations on a familiar theme. We already have analogous problems with equal access to high-quality education for children. And in the decades ahead we will increasingly have to worry about the technological illiteracy of people from deprived educational backgrounds. I do not mean to suggest these are unimportant concerns—far from it. Still, I am inclined to think that if problems of equity and discrimination were the only problems human genetic engineering generated, most people would welcome it as an almost unmixed blessing. Given the enormous increase in knowledge required to function in our increasingly technological society, it might well be argued that the capacity to improve our learning and reasoning abilities by genetic engineering had arrived just in the nick of time.

As we gradually map and learn to manipulate the human genome, however, it will become possible to alter or enhance many traits, not merely those, like intelligence and memory capacity, that are generally desirable and convey an obvious competitive advantage. It is a good bet that tastes, character traits, and other aspects of personality have a substantial genetic component. I do not think it is beyond the bounds of realistic possibility that in the next generation or two—and perhaps very much sooner—prospective parents will be able to choose from a library of genes in redesigning their own offspring. Nor is there any reason to suppose that all people or all societies will make the same choices. However, if different societies, or different groups within our own society, make systematically different choices for several generations, we may begin to see the genetic fragmentation of the human species. The divisions that separate cultural groups may come to include genetic differences so profound that members of different groups will no longer be interfertile.

The Nature of Humankind

The Western tradition of moral philosophy has left us unprepared to deal with the dilemmas posed by genetic engineering. Consider first the quandaries that arise when we try to think about such central ethical questions as the nature of the good or moral life, against the background of the emerging genetic engineering technology. From Socrates down to the present, just about everyone who has pondered the question of how men and women ought to live their lives has presupposed that human nature is in large measure fixed. Of course, profound disagreements have arisen about what human nature is like. However, the moral issue has always been conceived of as attempting to determine what sort of life a person should lead, given that a human being is a certain type of creature. This tradition leaves us radically unprepared to think about the questions forced on us by the prospect of human genetic engineering. Sometime within the next century, and perhaps much sooner than that, the age-old presumption of a more or less fixed human nature may begin to dissolve. It will no longer suffice to decide what constitutes the good or moral life for the sort of creature we happen to be; we shall also have to decide what sort of creature (or creatures!) humankind ought to *become*.

A worry of a rather different sort arises when we turn our attention to the processes of moral dialogue and the attempt to resolve ethical disputes. The Western philosophical tradition offers many views about the nature of rational moral dialogue and the quest for ethical agreement. But, I think, a common strand runs through just about all theories on this subject. In one way or another the notion of a shared human nature is rung in to explain how it is possible for people to reach a meeting of the minds on moral matters. When we are able to transcend our cultural and ideological differences, and agree on some ethical principle or judgment, it is because, despite our manifest differences, we share our humanity in common. However, human genetic engineering threatens to undermine the foundations of rational ethical dialogue by fragmenting our common nature along social and ideological lines. How shall I reason with a Moslem fundamentalist or a Marxist or a Moonie if what divides him from me is not merely his traditions and his convictions, but also his genetics? The prospect is at once so staggering and so unprecedented that we hardly know how to begin thinking about it.

Many people are inclined to think that the proper course of action is to put on the brakes in an effort to avoid ever reaching the point where these hard choices will have to be made. Many find it deeply distressing that we should even contemplate significant alterations in the human genome. Others worry that the power quite literally to remake our species is a power humankind will not use wisely, and because of this worry, they urge that we take steps now to ensure that this power will not be acquired. I have considerable sympathy with some of the concerns that underlie the recommendation to put on the brakes, though that is not the recommendation I would make.

I think two rather different arguments have led people to think we ought not to acquire the capacity to manipulate the human genome. One of these arguments is theological. God, it is said, designed humans as He wished them to be, and humans alter God's plan at their peril. Now, I am no theologian, but it seems to me this argument should evoke deep skepticism. For even if we grant for the sake of argument that God has a design or plan for the human species, we must take account of the overwhelming weight of evidence indicating He chose to unfold this plan via the mechanisms of evolution and natural selection. Species are not fixed over time, and each of us alive today had distant ancestors who were far more genetically different from us than we are likely to be from any imaginable genetically engineered descendants. The fallacy of the theological argument is to equate the divine plan with the status quo. Since the Renaissance, this sort of argument has been used repeatedly in an effort to oppose technological or social innovations that threatened to have a major impact on the structure of societies and the way people live. None of us, I would venture, are tempted to think that the technological, social, and economic patterns of the late Middle Ages reflected God's plan for how we should live. Nor do we think anything is sacrosanct about the genetic endowment of our Cro-Magnon forebears. I see no more reason to think God's ultimate genetic plan—if He has one—is reflected in the genetic composition of late-twentieth-century humankind.

Fiddling with the Gene Pool

The second argument against acquiring the capacity to manipulate the human genome rests not on a theological premise but on a scientific one. The current constitution of the human gene pool, it is argued, is no accident. We became the sort of creatures we are as the result of millions of years of natural selection. During those millions of years, many genes disappeared from the gene pool because the characteristics they impart to the organism were less adaptive than the surviving alternatives. Thus in a sense the current genetic makeup of humankind stores a great treasury of information about the sort of design that can flourish in our environment. It is folly, this argument concludes, to fiddle with the hard-won "evolutionary wisdom" bequeathed to us in our gene pool.

Now, although I ultimately disagree with this argument, I have considerable sympathy with the central insight it is urging. Human beings—indeed all currently existing species—are highly evolved, extraordinarily complex, and marvelously well-adapted to their natural ecology. We should be very wary indeed about altering components of this system until we have a good understanding of what role the components play in the overall organization of the system. However, it is one thing to suggest we act cautiously, keeping in mind that there are generally good evolutionary reasons for an organism's genome being the way it is. It is quite another thing to suppose natural selection cannot be improved on. To accept that view is to accept the Panglossian [excessively optimistic] assumption that the status quo is the best of all possible worlds. And that assumption is simply not true. A single example may serve to make the point. It now appears that the gene for sickle cell disease survived and flourished because an individual who carries only one such gene is better able to cope with malaria. In an area where malaria is endemic, the sickle cell gene conveys a selective advantage. However, when the swamps are drained and the mosquito population declines, the sickle cell gene is no longer worth having. What this example illustrates is that genes that may have been useful in the environment in which the species evolved may cease to be adaptive when the species finds itself in a new environment. But, of course, the environment in which humankind now exists is radically different from the environment that shaped the genome of our hunter-gatherer forebears. Thus we have every reason to think that the results of natural selection *can* be improved on.

Education Is the Key

Let me close with some brief observations on the policy implications of these reflections. As I have already indicated, I think it would be a serious mistake to adopt policies aimed at preventing the development of a technology capable of making major modifications in the human genome. However, it is certainly an area that cries out for ongoing, informed monitoring. Thus I endorse with enthusiasm proposals for an independent body, made up of scientists, ethicists, religious leaders, educators, and lay people whose function would not be to regulate but to study issues as they appear on the horizon. However, I am inclined to urge an even stronger role for education in dealing with the challenge of genetic engineering and other new technologies. Unless the public and the political leaders who represent them come to have a better understanding of the basic science underlying these new technologies, we have little hope that our social decisions will be wise ones. This understanding does not come easy, and it will be expensive. But in the long run the distressingly low level of scientific understanding in our society will be more expensive still.

Periodical Bibliography

The following articles have been selected to supplement the diverse views presented in this chapter.

John C. Fletcher and W. French Anderson
: "Germ-Line Gene Therapy: A New Stage of Debate," *Journal of Law, Medicine & Ethics*, Spring/Summer 1992. Available from 765 Commonwealth Ave., Suite 1634, Boston, MA 02215.

Kenneth A. Giles
: Testimony before the U.S. House of Representatives Committee on Science, Space, and Technology, Subcommittee on Natural Resources, Agricultural Research, and Environment, May 5, 1988. Available from the Government Printing Office, 732 N. Capitol St. NW, Washington, DC 20401.

Malcolm Gladwell
: "Risk, Regulation, and Biotechnology," *The American Spectator*, January 1989.

David J. Glass
: "Regulating Biotech: A Case Study," *Forum for Applied Research and Public Policy*, Fall 1989. Available from the Energy, Environment, and Resources Center, University of Tennessee, Knoxville, TN 37996-0710.

Russ Hoyle
: "Biotech Needs an Industry/Government Initiative," *Bio/Technology*, April 1995.

Jay Katz
: "Do We Need Another Advisory Commission on Human Experimentation?" *Hastings Center Report*, January/February 1995. Available from 255 Elm Rd., Briarcliff Manor, NY 10510.

Sheldon Krimsky et al.
: "Controlling Risk in Biotech," *Technology Review*, July 1989.

Isabelle Meister
: "Genetic Engineering: Urgent Need for a Safety Protocol," *Our Planet*, vol. 6, no. 4, 1994. Available from PO Box 30552, Nairobi, Kenya.

Lisa J. Raines
: "The Mouse That Roared," *Issues in Science and Technology*, vol. 4, no. 4, Summer 1988.

David Schardt
: "Diving into the Gene Pool," *Nutrition Action Healthletter*, July/August 1994. Available from the Center for Science in the Public Interest, 1875 Connecticut Ave. NW, Suite 300, Washington, DC 20009-5728.

Glossary

allele   One of several possible forms of a **gene**, found at the same location on a **chromosome**, which can give rise to noticeable hereditary differences.

amniocentesis   A surgical procedure in which a syringe is used to draw a sample of the amniotic fluid surrounding the growing fetus; by examining this fluid, physicians can determine the gender and possible genetic abnormalities in the fetus.

assay   An analysis to determine the presence, absence, or quantity of one or more components.

biotechnology   Defined a variety of ways, in its broadest sense it includes **genetic engineering**, reproductive engineering, and the use of genetic technology to address environmental problems and to develop agricultural products; some apply the term exclusively to the use of genetic technology for agricultural purposes.

bovine somatotropin (BST); also called bovine growth hormone (BGH)   A genetically engineered growth hormone used to increase milk production in dairy cows.

carrier   A person carrying a particular **gene** within his or her **genome** where it remains essentially inactive; the gene may be passed on to the carrier's offspring in which, because of the presence of another gene contributed by the other parent, it may express itself as a trait.

chimera   An organism with a mixed genetic heritage, formed by the merging of cells from two or more embryos.

chorionic villus sampling (CVS)   A prenatal genetic test, usually done at about eleven weeks gestation, in which a sample of the chorion (the fetus's outer membrane) is taken and tested to determine the presence or absence of genetic abnormalities.

chromosome   A chain of genetic material in the cell nucleus, consisting of **DNA**, **RNA**, and protein.

clone   An exact replica of a **gene**, cell, bacterium, or other organism; also, to create such replicas.

DNA   Deoxyribonucleic acid; the genetic material found in all living things; it exists in cells in the form of a **double helix**.

DNA fingerprinting   A set of techniques in which blood, hair, saliva, or other body tissue is examined in order to identify the owner based on the unique sequences found in his or her **DNA**.

dominant   A genetic trait that is apparent even if only one parent contributes the **allele** associated with it.

double helix   The geometric configuration of **DNA**, consisting of two complementary strands, each made up of a long, repeating sequence of sugar and phosphate molecules, running side by side in a spiral formation, and joined by certain chemical bonds.

enzyme   A protein that promotes or speeds up a chemical reaction in a cell.

eugenics   The science of improving a race or breed through some form of genetic control, such as selective breeding.

gamete   A reproductive cell, such as an egg, sperm, or pollen grain.

gene   A specialized segment of **DNA** whose sequence encodes the structure of a protein; genes are responsible for all the inherited characteristics of all life-forms.

gene mapping   Finding the locations on **chromosomes** of specific **genes**.

gene pool   All the **genes** in a certain population.

gene splicing   Any of various methods by which **recombinant DNA** is produced and made to function in an organism.

gene therapy   The insertion of normal or altered **genes** into cells in an attempt to overcome the effects of defective genes that cause disease.

genetic code   The sequence of nitrogen bases in **DNA** that form instructions for producing proteins.

genetic counseling   Counseling provided by trained professionals who inform clients about the implications of their biological family's health history or of genetic tests that can identify genetic abnormalities in themselves or their children.

genetic diversity   The variety of **genes** within a species.

genetic engineering   A technology used to alter the genetic material of living cells so that they will produce new substances or perform new functions.

genetics   The branch of biology dealing with heredity and variation in organisms.

genetic screening   The testing of individuals and populations for genetic abnormalities.

gene transfer   The introduction of a foreign **gene** into an organism.

genome   The complete set of **genes** in an organism.

genotype   An organism's genetic makeup.

genus, plural **genera**   A category of biological classification falling between the family and the species.

germ cell   A reproductive cell.

germ-line gene therapy   The insertion of normal **genes** into a **gamete** or fertilized egg in an attempt either to correct a genetic defect or to improve the **genome**.

heredity   The transmission of characteristics from parents to offspring.

hormone   A substance secreted by one type of cell that carries a signal to influence the activity of another type of cell.

Human Genome Project   The federally funded initiative to map and sequence the entire three billion base pairs of the human **genome**.

human growth hormone   A **hormone** secreted by the pituitary gland that stimulates growth of bones and muscles.

intergeneric   Existing or occurring between **genera**.

marker gene   A **gene** used to help recognize and identify other genes or gene patterns.

222

Mendel's laws   The findings and conclusions of Gregor Johann Mendel (1822–1884), an Austrian monk whose experiments on peas led him to discover that an inherited characteristic is determined by the combination of two hereditary units (**genes**), one from each of the parental reproductive cells.

muscular dystrophy   A hereditary disease characterized by progressive wasting of muscles.

mutation   A permanent alteration in a **gene** or **DNA** molecule.

nucleus   The part of a cell that contains the **chromosomes** and the bulk of the cell's **DNA**.

phenotype   The outward, physical manifestation of an organism.

polymer   A substance of high molecular weight, made up of a chain of identical repeating base units.

polymerase chain reaction (PCR)   A laboratory procedure in which **enzymes** are used to replicate a tiny amount of **DNA** over and over until the sample is sufficiently large for chemical analysis or experimentation.

population genetics   A field of **genetics** that focuses on breeding methods, **mutations**, and other factors to learn about genetic diseases and the frequency of certain **genes** within populations.

probe   A substance used by researchers to locate and identify particular segments of **DNA**.

recessive   A genetic trait that cannot become noticeable unless both parents contribute the **allele** associated with it.

recombinant DNA   DNA that is created in the laboratory by combining DNA fragments from different organisms.

restriction enzyme; also called restriction enonuclease   Any of various **enzymes** that break double-stranded **DNA** into fragments at specific sites in the interior of the molecule.

RNA   Ribonucleic acid; molecules made from and closely resembling **DNA**; these molecules carry genetic messages from DNA to the rest of the cell.

sex chromosomes   The X and Y **chromosomes**; women have two X chromosomes, while men have one X and one Y.

sex-linked   An inherited trait determined by a **gene** on a sex **chromosome**.

somatic cell   A body cell not involved in reproduction.

transgenic   A plant or animal into which has been inserted **DNA** from another species.

virus   An organism made up of a protein coat and an inner core of genetic material; it can only reproduce inside cells of other organisms.

X chromosome   *See* **sex chromosomes**.

Y chromosome   *See* **sex chromosomes**.

zygote   A fertilized egg.

For Further Discussion

Chapter 1

1. B. Julie Johnson expresses concern that scientists may lose control of the bioengineered organisms they create. How does Bernard D. Davis respond to such concerns? Whose argument is more convincing, and why?

2. Johnson uses the term *biotechnocrat* to describe members of the biotechnology industry. Does this word have positive or negative connotations? Does Johnson's use of this term make her argument more or less persuasive? Explain your answer.

3. Richard J. Mahoney is chairman and chief executive officer of Monsanto Company, a chemical company that relies on biotechnology research. Brian Tokar is an activist for environmental and social issues. How are the backgrounds of these two authors evident in their views about the safety and regulation of biotechnology? Which of their viewpoints is more convincing, and why? Does knowing their backgrounds influence your assessment of their arguments? Explain your answer.

Chapter 2

1. According to John Dyson, genetic engineering will enable the agriculture industry to provide consumers with improved products. Joel Keehn argues that biotechnology will benefit large agricultural and chemical companies but could pose risks to small farmers and the environment. Which author makes a stronger case? Support your answer with examples from the viewpoints.

2. What does Wes Jackson mean when he writes that agriculture should "integrate ethics, poetry, and spiritual dimensions into our fields"? Based on your reading of this chapter, do you agree with Jackson? Why or why not?

3. Andrew Kimbrell describes "pig No. 6707" to illustrate his concern about the suffering experienced by genetically engineered animals. Duane C. Kraemer addresses the issue of how animals should be treated by describing the diversity of opinions on the subject. Which author's technique better supports his position on genetic engineering? Explain your answer.

Chapter 3

1. Ruth Hubbard and Elijah Wald argue that DNA fingerprinting is unreliable in part because DNA samples can be con-

taminated. How does William Tucker respond to this concern? Does Tucker's explanation convince you that DNA fingerprinting is dependable? Why or why not?

2. Pat Spallone cites numerous problems with the use of DNA fingerprinting for paternity testing. Do you agree with her concerns? Why or why not?

Chapter 4

1. Geneticists are attempting to identify specific genes that are associated with certain diseases so that they can develop treatments and cures for these ailments. Ruth Hubbard and Elijah Wald argue that focusing on genetic "causes" ignores the social and economic factors that contribute to illness. Based on the viewpoints in this chapter, do you think medical researchers' emphasis on genes is appropriate or misplaced? Support your answer with examples from the viewpoints.

2. After explaining that eugenics was part of the philosophy of the Nazis of Germany, who exterminated millions of people during World War II, Pat Spallone describes gene therapy as "the new eugenics." Does her reference to the Nazis and her use of the term "eugenics" strengthen or weaken her argument? Explain your answer.

Chapter 5

1. Geoffrey Baskerville argues that human genetic engineering should be regulated by the federal government. Stephen P. Stich contends that it should not be regulated but rather monitored by scientists, ethicists, religious leaders, and others. Which author do you agree with, and why?

2. Those who favor extensive regulation of genetic engineering often base their opinions on their belief that the technology poses threats to human health and the environment. Based on the viewpoints in this book, do you believe genetic engineering is dangerous? Illustrate your answer with examples from the viewpoints.

Organizations to Contact

The editors have compiled the following list of organizations concerned with the issues debated in this book. The descriptions are derived from materials provided by the organizations. All have publications or information available for interested readers. The list was compiled on the date of publication of the present volume; names, addresses, and phone numbers may change. Be aware that many organizations take several weeks or longer to respond to inquiries, so allow as much time as possible.

Ag Bioethics Forum
115 Morrill Hall
Iowa State University
Ames, IA 50011
(515) 294-4111

The forum is an interdisciplinary group that focuses on the relationship between agriculture and bioethics. Among other issues, it explores the ethical dilemmas that arise when genetic engineering is applied to agriculture. The forum publishes the newsletter *Ag Bioethics Forum*.

American Civil Liberties Union (ACLU)
132 W. 43rd St.
New York, NY 10036
(212) 944-9800

The ACLU champions the civil rights provided by the U.S. Constitution. It is becoming increasingly concerned with the effects of genetic engineering on the right to privacy and the rights of defendants in criminal trials. The ACLU publishes a variety of handbooks, pamphlets, reports, and newsletters, including the quarterly *Civil Liberties* and the monthly *Civil Liberties Alert*.

American Society of Law, Medicine, and Ethics
765 Commonwealth Ave., Suite 1634
Boston, MA 02215
(617) 262-4990
fax: (617) 437-7596

The society's members include physicians, attorneys, health care administrators, and others interested in the relationship between law, medicine, and ethics. It takes no positions but acts as a forum for discussion of issues such as genetic engineering. The organization has an information clearinghouse and a library. It publishes the quarterlies *American Journal of Law and Medicine* and *Journal of Law, Medicine, and Ethics*; the periodic *ASLM Briefings*; and books.

B.C. Biotechnology Alliance (BCBA)
1122 Mainland St., #450
Vancouver, BC V6B 5L1
CANADA
(604) 689-5602
fax: (604) 689-5603
Web site: http://www.biotech.bc.ca/bcba/

BCBA is a nonprofit trade association for producers and users of bio-technology. The alliance works to increase public awareness and understanding of biotechnology, including the awareness of its potential contributions to society. The alliance's publications include the bi-monthly newsletter *Biofax* and the annual *Directory of BC Biotechnology Capabilities*.

Biotechnology Industry Organization (BIO)
1625 K St. NW, #1100
Washington, DC 20006
(202) 857-0244
fax: (202) 857-0237

BIO is composed of companies engaged in industrial biotechnology. It monitors government actions that affect biotechnology and through its educational activities and workshops promotes increased public understanding of biotechnology. Its publications include the bimonthly newsletter *BIO Bulletin*, the periodic *BIO News*, and the book *Biotech for All*.

Council for Responsible Genetics
5 Upland Rd., Suite 3
Cambridge, MA 02140
(617) 868-0870
fax: (617) 864-5164

The council is a national organization of scientists, health professionals, trade unionists, women's health activists, and others who want to make sure that biotechnology is developed safely and in the public interest. The council publishes the bimonthly newsletter *GeneWatch* and position papers on the Human Genome Initiative, genetic discrimination, germ-line modifications, and DNA-based identification systems.

Environmental Protection Agency (EPA)
401 M St. SW
Washington, DC 20460
(202) 260-4700

The EPA administers federal environmental policies, conducts research, enforces regulations, and provides information on many environmental subjects, including biotechnology. Its Pesticides and Toxic Substances division studies how the bacterial by-products and industrial chemicals produced by biotechnology affect the environment. The agency publishes the *EPA Journal* and many other publications on bio-technology and the environment.

Foundation on Economic Trends
1130 17th St. NW, #630
Washington, DC 20036
(202) 466-2823
fax: (202) 429-9602

The foundation examines the environmental, economic, and social consequences of genetic engineering. It believes society should use extreme caution in implementing genetic technologies because it fears that the unwise use of these technologies threatens people, animals, and the environment. The foundation publishes the books *Biological Warfare: Deliberate Release of Microorganisms* and *Reproductive Technology* as well as articles and research reports.

Friends of the Earth
218 D St. SE
Washington, DC 20003
(202) 544-2600
fax: (202) 543-4710

Friends of the Earth monitors legislation and regulations that affect the environment. It speaks out against what it perceives as the negative impact biotechnology can have on farming, food production, genetic resources, and the environment. Friends of the Earth publishes the quarterly newsletter *Atmosphere* and the magazine *Friends of the Earth/Not Man Apart* ten times a year.

Genetics Society of America
9650 Rockville Pike
Bethesda, MD 20814
(301) 571-1825
fax: (301) 530-7079

The society promotes professional cooperation among persons working in genetics and related sciences. It publishes the monthly journal *Genetics*.

Health Resources and Services Administration
Department of Health and Human Services
Genetic Services
5600 Fishers Lane
Rockville, MD 20857
(301) 443-1080
fax: (301) 443-4842

The administration provides funds to help develop or enhance regional and state genetic screening, diagnostic, counseling, and follow-up programs. It provides funds to develop community-based psychological and social services for adolescents with genetic disorders. It has many publications through its educational programs, and it produces directories and bibliographies on human genetics.

Kennedy Institute of Ethics
Georgetown University
1437 37th St. NW
Washington, DC 20057
(202) 687-8099
library: (800) 633-3849
fax: (202) 687-6779

The institute sponsors research on medical ethics, including ethical issues surrounding the use of recombinant DNA and human gene therapy. It supplies the National Library of Medicine with an online database on bioethics and publishes an annual bibliography in addition to reports and articles on specific issues concerning medical ethics.

March of Dimes Birth Defects Foundation
1901 L St. NW, #260
Washington, DC 20036
(202) 659-1800
fax: (202) 296-2964

The March of Dimes is concerned with preventing and treating birth defects, including those caused by genetic abnormalities. It monitors legislation and regulations that affect health care and research; awards grants for research; provides funding for treatment of birth defects; offers information on a wide variety of genetic diseases and their treatments; and publishes the quarterly newsletter *Genetics in Practice*.

National Institutes of Health (NIH)
Health and Human Services Dept.
Human Genome Research
9000 Rockville Pike
Bethesda, MD 20892
(301) 402-0911
fax: (301) 402-0837

The NIH plans, coordinates, and reviews the progress of the Human Genome Project and works to improve techniques for cloning, storing, and handling DNA. It offers a variety of information on the Human Genome Project.

U.S. Department of Agriculture
Grants and Program Systems
901 D St. SW
Washington, DC 20250
(202) 401-1761
fax: (202) 401-6488

The Grants and Program Systems division of the U.S. Department of Agriculture administers grants for biotechnology research and oversees such research. It has numerous publications on agriculture and biotechnology.

Bibliography of Books

George J. Annas and Sherman Elias, eds. — *Gene Mapping: Using Law and Ethics as Guides.* New York: Oxford University Press, 1992.

Eric Brunner — *Bovine Somatotropin: A Product in Search of a Market.* London: London Food Commission, 1988.

Iver P. Cooper — *Biotechnology and the Law.* New York: Clark Boardman Co., 1989.

Elaine Draper — *Risky Business: Genetic Testing and Exclusionary Practices in the Hazardous Workplace.* New York: Cambridge University Press, 1991.

Troy Duster — *Backdoor to Eugenics.* New York: Routledge, 1990.

Sherman Elias and George J. Annas — *Reproductive Genetics and the Law.* Chicago: Year Book Medical Publishers, 1987.

Federal Coordinating Council for Science, Engineering, and Technology — *Biotechnology for the Twenty-first Century.* Washington, DC: U.S. Government Printing Office, 1992.

J. Fiksel and V.T. Covello, eds. — *Safety Assurance for Environmental Introductions of Genetically Engineered Organisms.* New York: Springer-Verlag, 1988.

Michael W. Fox — *Superpigs and Wondercorn: The Brave New World of Biotechnology . . . and Where It All May Lead.* New York: Lyons & Burford, 1992.

Steven M. Gendel et al., eds. — *Agricultural Bioethics.* Ames: Iowa State University Press, 1990.

D. Goodman, B. Sorj, and J. Wilkinson — *From Farming to Biotechnology: A Theory of Agro-Industrial Development.* London: Basil Blackwell, 1987.

Joan Dye Gussow — *Chicken Little, Tomato Sauce, and Agriculture: Who Will Produce Tomorrow's Food?* New York: Bootstrap Press, 1991.

Stephen S. Hall — *Invisible Frontier: The Race to Synthesize a Human Gene.* London: Sidgwick & Jackson, 1988.

Chuck Hassebrook and Gabriel Hegyes — *Choices for the Heartland: Alternative Directions in Biotechnology and Implications for Family Farming, Rural Communities, and the Environment.* Ames: Iowa State University Press, 1989.

Henk Hobbelink	*Biotechnology and the Future of World Agriculture.* London: Zed Books, 1991.
Neil A. Holtzman	*Proceed with Caution: Predicting Genetic Risks in the Recombinant DNA Era.* Baltimore: Johns Hopkins University Press, 1989.
Ruth Hubbard and Elijah Wald	*Exploding the Gene Myth.* Boston: Beacon Press, 1993.
Stephanie Jones	*The Biotechnologists.* New York: Macmillan, 1992.
Steve Jones	*The Language of Genes: Solving the Mysteries of Our Genetic Past, Present, and Future.* New York: Doubleday, 1993.
Calestous Juma	*The Gene Hunters: Biotechnology and the Scramble for Seeds.* Princeton, NJ: Princeton University Press, 1989.
Daniel J. Kevles and Leroy Hood, eds.	*The Code of Codes: Scientific and Social Issues in the Human Genome Project.* Cambridge, MA: Harvard University Press, 1992.
Andrew Kimbrell	*The Human Body Shop: The Engineering and Marketing of Life.* New York: HarperCollins, 1993.
Jack R. Kloppenburg Jr.	*First the Seed: The Political Economy of Plant Biotechnology.* Cambridge: Cambridge University Press, 1988.
Sheldon Krimsky	*Biotechnics and Society: The Rise of Industrial Genetics.* New York: Praeger, 1991.
Thomas F. Lee	*Gene Future: The Promise and Perils of the New Biology.* New York: Plenum Press, 1993.
Thomas F. Lee	*The Human Genome Project: Cracking the Genetic Code of Life.* New York: Plenum Press, 1991.
Richard M. Lerner	*Final Solutions: Biology, Prejudice, and Genocide.* University Park: Pennsylvania State University Press, 1992.
Joseph Levine and David Suzuki	*The Secret of Life: Redesigning the Living World.* Boston: WGBH Educational Foundation, 1993.
Richard Lewontin, Steven Rose, and Leon J. Kamin	*Not in Our Genes: Biology, Ideology, and Human Nature.* New York: Penguin, 1984.
Robert J. Lifton	*The Nazi Doctors.* New York: Basic Books, 1986.
Jean L. Marx	*A Revolution in Biotechnology.* Cambridge: Cambridge University Press, 1989.
Marque-Luisa Miringoff	*The Social Costs of Genetic Welfare.* New Brunswick, NJ: Rutgers University Press, 1991.

Joseph J. Molnar and Henry Kinnucan, eds. *Biotechnology and the New Agricultural Revolution.* Boulder, CO: Westview Press, 1989.

Benno Muller-Hill *Murderous Science.* New York: Oxford University Press, 1988.

Dorothy Nelkin and Laurence Tancredi *Dangerous Diagnostics: The Social Power of Biological Information.* New York: Basic Books, 1989.

G.J. Persley *Beyond Mendel's Garden: Biotechnology in the Service of World Agriculture.* Washington, DC: CAB International and the World Bank, 1990.

Donald Plucknett et al. *Gene Banks and the World's Food.* Princeton, NJ: Princeton University Press, 1987.

Jeremy Rifkin *Algeny.* New York: Viking, 1983.

Barbara Katz Rothman *The Tentative Pregnancy: Prenatal Diagnosis and the Future of Motherhood.* New York: Penguin, 1986.

Robert Shapiro *The Human Blueprint: The Race to Unlock the Secrets of Our Genetic Script.* New York: St. Martin's Press, 1991.

Pat Spallone *Generation Games: Genetic Engineering and the Future for Our Lives.* Philadelphia: Temple University Press, 1992.

David Suzuki *Genetics: The Clash Between the New Genetics and Human Values.* Cambridge, MA: Harvard University Press, 1989.

David Suzuki and Peter Knudtson *Genethics.* Cambridge, MA: Harvard University Press, 1989.

Robert Teitelman *Gene Dreams.* New York: Basic Books, 1989.

Larry Thompson *Correcting the Code: Inventing the Genetic Cure for the Human Body.* New York: Simon & Schuster, 1994.

U.S. Congress, Office of Technology Assessment *Biotechnology in a Global Economy.* Washington, DC: U.S. Government Printing Office, 1991.

U.S. Congress, Office of Technology Assessment *Genetic Monitoring and Screening in the Workplace.* Washington, DC: U.S. Government Printing Office, 1990.

U.S. Congress, Office of Technology Assessment *Genetic Witness: Forensic Uses of DNA Tests.* Washington, DC: U.S. Government Printing Office, 1990.

U.S. Congress, Office of Technology Assessment *Mapping Our Genes: Genome Projects—How Big? How Fast?* Washington, DC: U.S. Government Printing Office, 1988.

U.S. Congress, Office of Technology Assessment	*Medical Testing and Health Insurance.* Washington, DC: U.S. Government Printing Office, 1988.
James D. Watson	*The Double Helix.* New York: Atheneum, 1978.
Peter Wheale and Ruth McNally	*Genetic Engineering: Catastrophe or Utopia?* New York: St. Martin's Press, 1988.
Peter Wheale and Ruth McNally, eds.	*The BioRevolution: Cornucopia or Pandora's Box?* London: Pluto Press, 1990.
World Bank	*Agricultural Biotechnology: The Next "Green" Revolution?* Washington, DC: World Bank, 1990.

Index

235